A Quiet Revolution

One District's Story of Radical Curricular Change in High School Mathematics

A Quiet Revolution

One District's Story of Radical Curricular Change in High School Mathematics

by

Michael D. Steele
University of Wisconsin-Milwaukee

Craig Huhn
Holt (MI) High School

INFORMATION AGE PUBLISHING, INC.
Charlotte, NC • www.infoagepub.com

Library of Congress Cataloging-in-Publication Data

CIP record for this book is available from the Library of Congress
http://www.loc.gov

ISBNs: 978-1-64113-181-0 (Paperback)

        978-1-64113-182-7 (Hardcover)

        978-1-64113-183-4 (ebook)

CONTENTS

PREFACE

Daniel Chazan

It is a pleasure to be invited to write this preface and to have recently had the chance to have a short visit at Holt High School (in what for me is the "new" building). Reading these chapters made me think back to interviewing at Michigan State University (MSU) in the spring of 1990 and, as a part of that interview process, being taken by Perry Lanier to meet his collaborators at Holt High School. Holt High School was then newly a Professional Development School. In particular, we met with Bill York and other members of the Mathematics Department. Though their collaboration was just a few years old at that point, there was already energy and a willingness to consider innovations, like having a Michigan State faculty member teach one period a day at the school. I left feeling that this was a unique context for collaborative work.

Those recollections make the stories related in this book all the more meaningful for me; they suggest that it is important to keep learning from a setting in which there is such a rare, long-standing, and deep relationship between school and university. For example, Holt High School has been a part of the MSU teacher education effort for a long time; secondary mathematics interns (as well as interns in other content areas) have been learning to teach from faculty in the Holt Mathematics Department for over 30 years.

And, of course, there has been much more to this relationship than only teacher education efforts; Holt teachers and their collaborators examine mathematics teaching and work to improve it within their building as well as more broadly in the state. It has been fascinating to read about how the teaching and learning of mathematics at Holt High School have continued to develop and change through shifts in policy context over the last slightly more than a decade. For example, this book documents how, as a part of the larger effort to change mathematics education in the United States, at Holt High School, there have been shifts in expectations for students; graduates of the school have gone from not needing to study "college bound" mathematics like Algebra 1, from only needing to studying arithmetic in the late 80s, to the expectation currently met at Holt that every student take and pass a mathematics class whose content would be labeled as Precalculus in other school contexts and which is labeled Algebra 2 at Holt. A quiet revolution, indeed! This is a compelling story that is worthy of attention given our country's decades-long efforts to help high school graduates be prepared to choose to go to college in general and to pursue STEM careers in particular.

As documented in the coming pages, the Holt High School Mathematics Department, as a professional context, continues to afford unique learning opportunities for Holt faculty members and for inducting new teachers into an important image of professionalism. Coming back to visit Holt from a different context where it feels like there is much less stability and greater flux in the work lives of teachers and therefore fewer rich learning opportunities, it was wonderful to meet with a number of teachers who I remember as MSU interns and who have now been in the profession and working together for almost 20 years; it was wonderful to read about how at Holt High School teaching continues to be a life-long career opportunity with supports for professional growth, like participation in the writing of this volume, as well as other opportunities for engaging mathematics education stakeholders outside of their building. It was wonderful to once again be immersed in an environment that treats the teaching of mathematics seriously as a long-term matter of great importance and to engage with teachers about the policy context in which their work plays out. The experiences and expertise that are evident in the rest of this volume are now being offered to a larger professional audience with suggestions for making similar professional engagement available in other sites.

To be sure, there are many parts of this story that involve local characteristics that are unique. For example, Holt is a school district with just one high school (that lately enrolls about 1,800 students). It is much different than the school districts of about 150,000 student in my current local context that have more than 20 high schools most of which are larger than Holt High School. There are ways in which the size of a district change what it means for teachers to have the sorts of professional opportunities outlined in this book. The local economy shapes the story that is told in this book as well. The economic forces that shape high school students sense of their possibilities after graduation, as well as mathematics teachers professional experiences, are different across this country. And, Michigan State's ongoing role in supporting the efforts of Holt teachers is clearly an important part of the story. Not every high school has this sort of relationship with a teacher certification granting institution, but many more schools could have such a relationship, particularly if we as a country organize to respond to looming teacher shortages, particularly in STEM fields.

I do not believe the authors of this book would ever argue that their story has some magic bullet in it that will transform mathematics teaching in all high schools across the United States. Interested readers will need to work hard to analyze the offerings in this book and to figure out what is most applicable to their own contexts. As I read, and think about my own experiences at Holt High School, what stands out to me is a deep commitment to conversation across differences of perspective and opinion and to ongoing efforts to experiment, innovate, and improve. Building on the image of mathematics teaching at Holt, I hope that readers of this book engage deeply with it, argue with it, and draw their own conclusions.

INTRODUCTION

Craig Huhn

In college, when I was about to enter my fourth year, I had no idea I was about to become radicalized.

I had decided on teaching, and had enjoyed the generic education courses that pushed my thinking on the purpose of schools and issues of equity. I knew from a discussion-based part of my own sixth grade experience that being a teacher meant having the opportunity to engineer discussions among a group of students that could expand their brain and their world-view. In my mind, teaching history at the secondary level had the best opportunity for dissecting the moral dilemmas of our past (Should the U.S. have dropped the atomic bomb? Is the Electoral College a fair system? Have there been circumstances where assassinations or torture could be justified?). I saw myself as the architect of class discussions where students grappled with complex ideas and learned to listen to ideas from peers. But when push came to shove and I had to choose my major and minor, my still-teenage self shied away from a history major and all the papers and reading the coursework would entail. Knowing that I could still teach my minor, I selected history in that slot and considered something easier to make my major. In a fateful move, I chose mathematics, and subsequently began my path of collegiate courses to earn those credentials. What I did not know at the time was that the way the teacher preparation program at MSU was set up, the methods courses would focus almost exclu-

A Quiet Revolution:
One District's Story of Radical Curricular Change in High School Mathematics, pp. xi–xv
Copyright © 2018 by Information Age Publishing
All rights of reproduction in any form reserved.

sively on my major. What was at the time an oversight ended up igniting a lifelong passion for math ed.

I walked into those methods courses assuming that to be a quality math teacher, all I had to do was be personable, and explain math really well. Over the course of the next several months, I was challenged to think about what it means to learn math—what it means to understand mathematics in a way I was quickly learning I lacked, even with a degree from a major university in mathematics all but complete at that point. I was challenged to reconsider what I thought mathematics actually *was*. And throughout this fourth year, while I watched classmates struggle with being pushed in this way, I welcomed this shift in world-view. I loved the discussions we had, trying to debate the nature of infinity and countability; dissecting the words of Aristotle as he grappled with continuity. *Arguing* about math! Thinking about who knows what and how we come to know truths. And wondering how one truly comes to understand ideas.

In that year, I reconceptualized for myself what math ed can look like. And going into the fifth year of the program, where teacher candidates were immersed in a year-long student teaching internship, I found out I was placed at the local high school that we had become familiar with in our courses as a math department that was working to teach mathematics in the way we were learning to teach it. During that fifth year, we also took a second year of methods courses at the university where we continued the journey to make sense of what I was realizing was a really complicated endeavor. I was immersed in a professional culture as an intern at Holt High School and trained to become a revolutionary in my coursework at the university. In my mind, it was no longer good enough to strive to teach well (with this new idea of what that means), I also felt a moral imperative to make sure this idea found traction in classrooms all over.

Around this time, the National Council of Teachers of Mathematics (NCTM) published its Principles and Standards for School Mathematics. This document, NCTM's second attempt at a national policy document for teaching and learning mathematics, was intended to be a more moderate document—a treaty in the Math Wars that had been pitting conceptual understanding and procedural fluency against one another. Both the push and pushback were playing out in real time as I was figuring out how complex working with 150 teenagers can be. In a continuation of fate, a position opened up at Holt the next year, and I have spent the past 17 years still working to get as many teenagers to learn and understand as much math as possible. I consider myself one of the most fortunate people in the world to have been trained under one of the top teacher ed institutions, under some of the best math ed thinkers in the world, placed in a school that had a professional community with ties to that program, and been able to continue my career among some of the best educators I have ever known.

It was in this first year or so that one of my math ed instructors (one of the aforementioned best in the world), Dan Chazan, came to us to try and tell the story of how Holt High School and Michigan State had forged their alliance and how it had created a laboratory for working to improve mathematics education. It was my first year teaching when I became involved in that project and wrote for *Embracing Reason* (Chazan, Callis, & Lehman, 2008). In the years following, I have continued to try and extend the conversation about teaching and learning through other articles, chapters, presentations, workshops, or supervising interns. I have continued to attempt to engage students in understanding mathematics (with successes and failures in almost equal measure), and to fight for structures and systems that support our ability to do so (with failures seeming to outnumber successes). And I have continued to think about how *Embracing Reason* would sound if it had been written now instead of in the early 2000s.

Often people say change in education is slow, and I do not often disagree. But recognizing how different my job was 15 years ago, both local changes and the state and national landscape, makes a decade and a half feel like a century and a half. Today, there is little argument among the professional community about what quality mathematics education looks like. NCTM has published *Principles to Actions*, which outlines a detailed vision of what the teaching and learning of mathematics with understanding looks like in classrooms that make use of rigorous mathematics standards like Common Core. The Common Core makes it clear that certain practices, consistent with what we now know are best practices, are expected in math classrooms across the country. When I began teaching, Michigan did not even have a state set of standards in force. Since then, they have written and passed several versions up to and including the Common Core. A standardized state test went from one outside measure that teachers used to see how students were doing, to a high-stakes assessment of the quality of your school that determined what degree of sanctions the state would impose. And of course, that state assessment has changed no less than four times at least (although all iterations, from the MEAP to the ACT to the SAT, honored speed and surface-level procedural knowledge of tricks in multiple choice format). My state, hit hard by the recession of 2008 and exacerbated by political choices, has drastically underfunded education. And, purely for political choices, my state has also weakened propublic education laws, curtailed the ability for teachers to collectively bargain and stand up for students, and made several (unsuccessful, so far) attempts to enact vouchers. One such change was the enactment of "Schools-of-choice," which allowed parents to take their kids (and the state money that is dedicated to educating them) to any school they want, as long as they are willing to transport them there. As we will describe later, this has had a significant impact on Holt.

Since *Embracing Reason*, the cast of characters from Michigan State and our department have largely turned over. The political landscape and state expectations have changed. Movements like "formative assessment" and "growth-mindset" have evolved into critical pieces of pedagogy. The population Holt serves has become much more diverse. Curriculum has adapted, evolved, and adjusted. Grading practices have done the same. And yet, we still struggle to engage all students. We still question and evaluate the decisions we make, and push each other to continue to improve, but fail to reach all students. Our struggles are not new, even if some of the circumstances or nuances are. However messy it can be, we have at least gotten to a place (mostly fallen into it, as you will soon see), where we have an urban-fringe, large(ish) high school that requires a course equivalent to Precalculus for all kids to graduate, and gives every student the opportunity to take AP Calculus or AP Statistics if they so choose. And, the world did not implode.

How to Use This Book

The transformation of mathematics teaching and learning at Holt High School is special, but it is not magical. The mathematics teaching and learning outcomes we describe in this book arise from constant discussions, agreements, disagreements, evolutions, and revolutions amongst stakeholders at Holt High School. Our goal in writing this volume is to chronicle the most important aspects of those ongoing conversations and negotiations and to provide you, the reader, with concrete tools to instigate challenging discussions in your school and district that have the potential to lead to transformative change.

Each chapter describes an aspect of the Holt High School mathematics program over the years, ranging from demographics to course offerings to student outcomes. We describe both the changes that took place within Holt High School over the years and the local, state, and national contexts in which those changes were situated. Changing the culture and norms of your school involves persistence and resilience, as the pull of the status quo in education is strong. To support you in effecting change, this book aims to radicalize you. As a *radical mathematics educator*, you will have the data and tools to challenge assumptions, instigate difficult conversations about beliefs and practices, and inquire thoughtfully into your own mathematics teaching practice and how that practice supports student learning. The close of each chapter includes an Investigation and Reflection Activity that prompts you to look both inward at your own classroom, students, and teaching practice; and outward towards your colleagues, administration, and community. The Investigation and Reflection Activities should begin with written reflections based on the prompt, and follow with concrete actions to prompt meaningful discussion within your school. We provide a framing for each conversation in the form of questions to pose, activities to engage in, and data to collect with your colleagues.

We encourage you to keep a physical or digital *Radicalization Handbook* as a record of your reflections and conversations. Collecting a narrative history of your own journey and the discussions in your school, district, and community can be a powerful tool for identifying levers for change. We also encourage you to reach out to others across districts and states to gain support for engaging in these conversations. Together, we are mighty and we can change the world of mathematics education for all students.

Let's begin.

CHAPTER 1

THE QUIET REVOLUTION

Three Decades of Mathematics Education at Holt High School

One of the challenges of talking about your practice with others is making sure that by engaging in a larger conversation, you are not suggesting you have all the answers. The goals of this book are to share some of what we have tried and learned in one particular location, but also to share the struggles and challenges we still have. We still have a long way to go—clearly, as we all do—and the reflections, analyses, and activities in this book are one attempt at taking the next step toward better addressing pressing issues in high school mathematics teaching and learning. The second tough part about talking about practice is trying to ensure the story is generalizable; or rather, not disregarded as being inapplicable to other settings and locations like yours.

Our aim in sharing this particular story of Holt High School (HHS) is to try and talk specifically about what can be done in *any* setting to try and take small and not-so-small steps toward improving mathematics education. This comes with the caveat (the very large, significant one) that individualized variables are of critical importance, but are yet manageable. It is with this in mind that we now lay out the variables that exist around Holt High School, in the hope that we can initially quiet the inner voices that stubbornly claim, "well, not *my* kids..." Because, at its heart, HHS is a pretty large school in an urban-fringe setting, with many of the issues

A Quiet Revolution:
One District's Story of Radical Curricular Change in High School Mathematics, pp. 1–15
Copyright © 2018 by Information Age Publishing

that students face at such schools. In 2015–16, the district has about 38% of its 5,712 students designated as receiving free or reduced lunch consideration. White students comprise 65% of the student population. Special education services are provided for 13% of students. Fifty students in this year were homeless. And, like any large(ish) high school of 1,800 students, students are often faced with tough life choices and sometimes make the poor ones. I [Craig] have had former students that have overdosed on heroin. I have had former students that were shot and murdered. There are occasional fights at the school. There are students underperforming. And yet, by and large, HHS is filled with students who also make better choices: putting forth effort in class in ways they never have before, staying after school for extra help, becoming the first student in the family to go to college, or donating time or money for various charitable causes.

This duality is reflected in the community that, like other urban-border areas, schizophrenically coexists as a suburban entity and a continuation of the larger city. Visitors driving under the I-96 overpass would have no other clue that they have travelled from Lansing into Holt aside from a small city line marker. And while some problems that larger urban centers have do not stop at any overpass, Holt also has been named one of the "happiest" places in Michigan in 2015 (Jamison, 2015), and, with easy access to Michigan State University, Lansing Community College, and Lansing's business center, one of the top ten places in Michigan to own a home (Olsen, 2014). In many ways, HHS and the surrounding area look and feel a lot like many other places in America. But it is neither a small, rural school, nor is it a huge complex that serves an inner-city population with struggles like Detroit. The intent of the book is to talk about what we did, in our context at Holt, with the barriers we had, and how it played out; and then talk about ways in which aspects of the same process might be brought to your localized scenario. We believe (I think safely) that the same tenets—collaboration, professional growth, systematic supports—are universal. Those characteristics are no particular surprise, as they have been named and described for years in mathematics education policy and leadership documents from the National Council of Teachers of Mathematics [NCTM] (2000, 2014) and the National Council of Supervisors of Mathematics (2008, 2014). What we offer that is different from other texts that discuss collaboration, professional growth, and systemic supports is a set of honest, accessible starting places for you to instigate your own quiet revolution using the lessons we have learned in Holt. The details of how you operationalize these tenets, however, will be modified as you get the ball rolling at a starting point that makes sense for your school and district. As you think through some aspects of the change that needs to take place, maybe the sociopolitical infrastructure already exists in your school, and you can use these ideas as a blueprint to take what makes sense and implement, and to study how it works. Other changes may take some politicking to even drum up support, be it financial or philosophical.

Maybe a second needs to be taken here to expound on that last point. When dealing with the learning of mathematics, there continues to be what NCTM called in 2000 a "pervasive societal belief in North America that only some students are capable of learning mathematics" (NCTM, 2000, p. 12). This has lessened some in professional circles as the work from Jo Boaler (2016) and Carol Dweck (2006) on growth mindsets has made headway, but in working with math teachers across the country as recently as the summer of 2017, it was clear that many math teachers still cling on to this debilitating orientation. Add to that the unflinching faith that some people have, contrary to everything we now know about best practices, that students learn best when a knowledgeable other provides the demonstration and corrective practice, and you have an initial level of resistance that could exist within your own department. This is not to say that it will prevent you from making meaningful changes; simply that your approach to making these changes will need to be adapted to account for this variable. A department with like-minded individuals from top to bottom will move forward to the next hurdle more quickly than a department with a smaller but motivated group of like-minded thinkers, and quicker still than a department with fierce resistance and dissent within. It's just a matter of finding out where you are at in the long and evolving process of improvement, making strategic decisions about how to proceed, and realizing that even a large wall comes down with smart decisions about where to chip away at it.

If you are lucky enough to have a critical mass of math teachers in your building that "get it," then your focus simply is on a larger potential barrier. In our minds, there is a series of concentric circles of probable encumbrance (trying hard not to call on Dante here for a fitting analogy), starting with your department in the center, surrounded by the other adults in the building that interact with students often, have their best interests at heart, and yet have not been part of the conversation over the last 30 years about high-quality effective math teaching. Many of them, particularly science and technology teachers, were good at mathematics, and therefore can subtly (or not so subtly) undermine efforts to make changes to the traditional system in which they likely flourished. Others may have gotten to a successful point in their career in spite of mathematics (in their minds) and carry a grudge that comes across as trying to save students from having to experience math. We mention this not only because they have the power or influence to veto changes you might make, but also because they can be an invisible barrier in the effort to change the hearts and minds of students about math teaching and learning. And, in cases of counselors or case

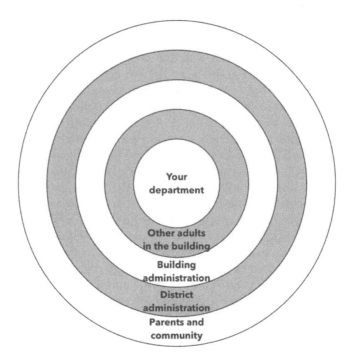

Figure 1.1.   Five circles of encumbrance.

managers, successful change hinges on the level to which they understand what you are trying to do in your classroom and trust you.

The third concentric circle represents building administrators which, in our experience, is the real secret. A building principal on board with an effective mathematics teaching approach is the single greatest thing that can make a difference for your students. Their decision-making power is clearly a necessary component to making and sustaining change, but their importance transcends that. Their support for your program and their belief in what you are doing has tendrils that extend back into your department and out into the further layers in ways that an impassioned, researched math teacher often does not. If you take nothing else from the book, take this—that sharing your thinking with your principal(s) as colleagues, particularly when you have no immediate agenda, is an investment in your students. Determining how to engage your principal in conversation depends on knowing them a bit as a person. The particulars will depend on whether your building leader is a new, young person coming in from outside the district, a lifer that has been around forever and hates you, or someone that existed before you got there and you know only as that person that comes in and comments on participation rather than substance. (These are broad thumbnail sketches of stances towards leadership, but we are sure you can think about where your principal sits relative to these portraits.) Secondary principals are an interesting bunch. They most often come from the secondary teaching corps, which means they have a depth of expertise in a particular content area that an elementary teacher turned principal, for example, may not have. Principals with this background also understand (at some level) that teaching mathematics is different than teaching English, science, social studies, music, and so on. It is likely that they have some specific pedagogical tools and resources in their subject area that were unique to that subject area—stances on how to support the development of writing skills in English, effective rehearsal techniques in music, design of inquiry-based labs in science. They may be years removed from this intensive work and thinking about effective content-based teaching, but that does not mean that you cannot hearken back to this work. Your overall goal is to help them understand what good teaching and learning looks like in mathematics, and that these practices are not personal views but grounded in research. Use their own teaching experiences as a launch pad for this work.

Next, understand where they are related to mathematics teaching and learning, what their own mathematical learning experiences are, and consider how you can use that information to help them make sense of your work and buy into it. Wherever they are at in the math education continuum at this point, from antireform to "gets it," your job is to find a leverage point and start to move them. Build trust—both personal trust and the trust that you are well informed about the work. Show them that you are incredibly thoughtful and knowledgeable. Lend them your copy of *Principles to Actions* (NCTM, 2014) and talk to them about their thoughts. Share articles with them. Help them understand that the ideas for which you are advocating are not your personal hippie crazy beliefs, but research-based, evidence-backed consensus about good mathematics teaching and learning. Connect to their own teaching work by asking about trends and developments in their subject area, and use that information to connect to the work that you are doing and where you wish for the department to go. The goal is not simply keeping them from being antiprogress, but feeling like they are committing malpractice by *not* changing—a true collaborative partner in making better mathematical experience for all the students in his or her charge.

The next circle is district administrators—superintendent, curriculum director(s), chief financial officer, special education director, and so forth. Often they reside outside of your building, and in larger districts they may even feel too far out of reach, but they are important influences. This group can have final say over course offerings, hiring and assigning instructional staff, financial allocations, and a variety of other things. These folks are the policymakers. The policy world may seem daunting—this administrative cohort is accountable to state and federal regulations and oversight. In most cases, however, the real work of policy is about implementation. For example, a state policy that mandates all students complete Algebra II to graduate could be operationalized in many ways, and most of those ways involve interactions between administration and teachers in classrooms who have the direct responsibility for the learning. Your administrative team does not have to understand the ins and outs of pedagogical approaches to social-constructivist learning theory in mathematics, they just have to trust you. And, more importantly, they listen to your building administrator for a lot of their decision making. As mentioned previously, that tendril is pretty important. But to the degree you can, finding an advocate and supporter in this ring is a sure bonus—many of them (particularly coaches or curriculum people) genuinely want to learn more about "the trenches" and you and your other patriots for change can be the ones to talk to them about quality math education.

And the final circle on the outside is the community. Parents, specifically. Individual math teachers who teach in student-centered ways have already honed their messages about what happens in their classroom and shared the rationales, and probably become pretty adept at bringing adults on board. In some communities, an overinvolved collusion of helicopter parents (probably spurred by social media's ability to foment malcontent and misinformation) can seem as powerful as a well-intentioned mafia. In others, the complete absence of parental involvement is its own challenge. Like the second ring, this one has potential to creep up and get in the way—you will have to consider in your world if you need to be proactive in some ways or if you can do your work while just keeping an awareness for questions that may not come to you initially. In either case, crafting a set of messages that you can use to explain what you do and engage parents and community stakeholders in conversation when necessary is important. The first step in creating such messages is spending some time considering your own beliefs and practices, and coming to terms with what you do and why you do it. The Investigation and Reflection Activity in this chapter will support you in thinking through how your beliefs intersect with your teaching practices, and crafting brief yet rich messages that will help you advocate for your work.

THE ANATOMY OF A REVOLUTION: AN OVERVIEW OF THE WORK IN HOLT

Over the past three decades, Holt High School has engaged in ongoing and continuous work aimed at strengthening mathematics outcomes for all students. The road to change has not been a straight line, and has not been without challenges and setbacks. In the chapters that follow, we detail the specific nature of the work at Holt and put those changes in the context of broader policy context. Table 1.1 summarizes the key shifts in each of the main areas that the remainder of the book describes, along with a brief description of the conversations that led to the changes, the evolution of the changes over time, and the challenges involved in making the change.

INVESTIGATION AND REFLECTION ACTIVITY 1: WHAT I DO AND WHY I DO IT

Often times when teachers are asked to consider their beliefs about mathematics teaching and learning, we ask them to list out their beliefs and then describe how the beliefs play out in their teaching practice. This is a fine endeavor, but it has one important caveat to it regarding mismatch. Our beliefs do not always match our practices—in fact, they almost never do in an exact way. The relationships between beliefs and actions is complex and nuanced, and change in each influences the other. Scholars have argued for decades about which is more important and where to begin (beliefs or actions) in supporting teacher change. We are sidestepping that argument entirely here.

Instead, we are going to focus on the piece of the beliefs-actions nexus that most directly influences student learning—your actions in the classroom. Create a table in your *Radicalization Handbook* like the one we have illustrated in Table 1.1, and identify a set of actions that take place regularly in the classroom (every day, every week, every unit…) that you feel are central to your work teaching mathematics. For each of those actions, list a belief about the teaching and learning of mathematics that relates to that practice. Your set of beliefs together may not be coherent—in fact, you may have some that seem to contradict one another. That's okay—and in fact, that is an expected outcome. None of us have a fully consistent set of beliefs.

Finally, identify a set of actions that you are not currently taking that you would like to take. For each of those actions, identify the circles of encumbrance to which the action most closely relates. It is easy to stay at the center—challenge yourself to identify actions you would like to take that relate to at least two different circles (see Action Table).

Once you have completed this activity, we include as a postscript to this chapter some perspectives on teaching and learning that serve as guiding principles for this book. These ideas reflect the current canon of research in mathematics education, are embodied in the NCTM Standards and Principles to Actions (NCTM, 2014), the Common Core State Standards [NGO & CCSSO] (National Governors Association Center for Best Practices, & Council of Chief State School Officers, 2010), and by and large the shared values of the HHS mathematics department.

Table 1.1.
A Overview of Holt's Key Shifts Across Three Decades

What's changed?	What led to the change?	What are the persistent challenges?
Demographics and Policy (Chapter 2)		
-shift from a more rural/urban fringe community to a more urban community	-population and geographic changes to the region over time	-being constantly vigilant and aware of student needs and affordances
-more diverse student demographics	-state policy changes to increase student mobility	-student mobility from systems that are not similar in structure
-increased student migration due to school choice	-state and national policy changes about graduation and standards	-using policy changes as productive catalysts for local reform
-changes to state policy increasing graduation requirements (4 years of math, including Algebra II)		
-moving from state standards to adoption of Common Core		
Ongoing evolution and implementation of the mathematics program (Chapters 3 and 4)		
-a long-term evolution of the 10th grade mathematics course from Algebra I content to Algebra II content	-a commitment to meaningful, context-rich mathematics teaching and learning	-maintaining and strengthening a student-centered approach to teaching and learning over time
-continuing to improve on a student-centered, discourse-based approach to teaching that represent a social constructivist perspective	-a collection of resources over time to support functions-based, context-rich teaching	-inducting new faculty into the work

-a meaning-based approach to functions as a foundational component of future mathematical study	-the promotion of a growth-mindset approach to mathematics among students	-moving the agenda forward productively through changes in administration, demographics, staff, and community
-de-tracking and de-stigmatizing repeating coursework	-a focus on course content and meaning making rather than course names	
	-valuing of mathematics education research and a close research-practice partnership grounded in work with Michigan State University	
	-consideration and discussion of long-term mathematical trajectories across courses within and outside HHS	
	-encouraging repeating content by removing punitive "no credit" course outcomes	

Student outcomes (Chapter 5)

-shifts in course offerings such that all students complete the equivalent of Precalculus	-shifting the focus from point-gathering to criteria-meeting in mathematics classrooms	-consistently balancing more procedural standardized assessments with conceptually-focused assessments
-opportunities for all students to complete AP Calculus or AP Statistics if they so choose	-just-in-time supports for student success (RAMS course)	-being vigilant about the success of each and every student
-more students taking more mathematics coursework over time	-placing student learning and local assessment in the foreground and large-scale assessment in the background	-using data to constantly evaluate successes and challenges and make programmatic adjustments
-relatively stable ACT scores measured against state averages	-strengthening students' self-image and beliefs with respect to mathematics learning	

(Table continues on next page)

Table 1.1.
(Continued)

What's changed?	What led to the change?	What are the persistent challenges?
Student outcomes (Chapter 5)		
-use of diverse and rich assessment measures, including oral exams and standards-based grading		
Shared vision and beliefs (Chapter 6)		
-regular work doing mathematics as a faculty together and talking about student learning of mathematics	-proactive communication with stakeholders about our approach to mathematics education and the outcomes	-shared vision and beliefs means selling students, parents, colleagues, and the community on the vision and research behind it
-helping other adults in the building to understand what is done and why	-exacting interview processes to determine candidates' dispositions and potential for development	-this work is best done simultaneously on multiple fronts and requires time and energy, particularly when proactive communication doesn't take place
-recruiting like-minded teachers when there are have faculty openings	-taking the time and space for mathematical work through late starts	

10

Actions That Are Core to My Classroom Practice	Beliefs That Undergird Those Actions
Actions I Would Like to Take Next	**Corresponding Circle of Encumbrance**

POSTSCRIPT: PERSPECTIVES ON TEACHING AND LEARNING THAT SERVE AS GUIDING PRINCIPLES

Perspective 1: Learning is Socially Constructed and Situated

A review of the historical evolution of learning theory is well beyond the scope of this volume. We include this brief passage to note the current perspective on learning and how it differs from historical conceptions, many of which undergird the structures of schooling today. Behaviorism, a theory of learning popular in the first half of the 20th century, posited that specific stimuli would elicit specific behaviors, which could lead to learning. The work of B.F. Skinner is the canonical example of behaviorist theory, and in particular his experiments asking rats to perform certain behaviors in exchange for food rewards. Behaviorism did not take into account any particular mechanisms of thought or will, instead suggesting that if one could identify the complex systems of stimuli that connected to a desired behavior, it was simply a matter of applying those stimuli accurately and consistently to elicit the target response. Punitive and extrinsic reward systems of discipline still rest on the behaviorist conception of learning, and many of our modern schooling structures were built on behaviorism.

In the mid-1970s, educational researchers and psychologists began to develop what we refer to as a cognitive perspective on learning. This work evolved the understanding of human thinking to explain how people constructed knowledge, the forms that knowledge might take in the brain, and how new learning relates to previous learning. The key components of the cognitive perspective are the development of schemata, networks of information that may be more or less robust. These schemata drive behavior in particular contexts. For example, most of us that have learned to drive a car have a schema for getting in and starting the car that includes unlocking the door, starting the car (by whatever mechanism your car accomplishes this task), checking the area before starting the car moving, positioning hands and feet in useful places, and so on. New learning occurs when we extend or adapt a schema to a new situation. (For example, when Mike's new car started via a push button rather than a key inserted into the steering column, Mike had to adapt his "start the car" schema. He still has trouble when he gets in the old car with the physical key.) Based on this work in cognitive science, the understanding of learning evolved from behaviorism to a constructivist perspective in the 80s and 90s.

The element that straight-ahead cognition does not take into account is the role that interactions play in learning. When two people talk about an idea, where does that knowledge live? Craig and Mike might have a conversation and take very different things away from it, which might change schemata that we have. But how does that learning take place? What catalyzed it, and where did it come from? Situated cognition and the sociocultural perspective on learning suggests that changes in knowledge do not come from interactions, but the interactions themselves are the knowledge. This recognition led to the understanding that learning always occurs in a context—there is no such thing as a completely abstracted generalization —because the only way that we can see if learning took place is to provide a context for which it can be demonstrated. Learning is shown by increasingly sophisticated performances on situations within a domain. This can be a radical way of thinking, particularly in a school-based setting that typically assumes that knowledge is a personal asset and values knowledge as an individual possession rather than as a shared construction. For much more on situated cognition, we cannot recommend highly enough the work of Jim Greeno (Greeno & the Middle School Mathematics through Application Project, 1998 is an excellent and very accessible starting point).

Perspective 2: Each and Every Student Can Learn Mathematics With Effort and Support

The ideas of equity, access, and empowerment have taken many forms over the past 30 years under many labels. We can use whatever we label we like to communicate that all students should be afforded meaningful opportunities to learn mathematics, but the work will not change outcomes for students unless teachers fervently believe the following: that all students, no matter what their previous experiences or background, can learn mathematics. Thanks to Carol Dweck (2006), we have a popular new phrase that partially describes this stance: *growth mindset*. The key idea behind the stance that each and every student can learn mathematics with effort and support is that our capacity to learn is not fixed or predetermined. Dweck contrasts growth mindset with fixed mindset, the notion that talent and ability in a particular area is predisposed, genetic, or somehow or another gifted to people in some way.

This fixed mindset belief is particularly pervasive in mathematics. Both Craig and Mike have heard many times, in many different contexts, from community members, parents, and even our own family members something like the following: "I was never any good at math either," or "I'm just not a math person." And while this rhetorical move has become a bit cliché, we do not say such things about reading and literacy, and we certainly do not position them as a point of pride in some contexts. If there is such thing as a math person, that suggests that there is such thing as not a math person, and people position this idea as a characteristic to be described, not a current state of being that can change.

Adopting a growth mindset means being vigilant about the messages we send to students through our interactions with them and the structural components of teaching and learning mathematics. This can be as small as calling a student "smart" in math class rather than being specific about what mathematically you thought was smart and why you as a teacher found it so. This can be as large as tracking systems that place labels on students early and make it hard or impossible for them to earn into a higher prestige label. This can be as insidious as other adults in the building subtly steering students towards or away from STEM careers based on a perception that they are or are not mathematics "people" without engaging them about their interests and goals.

It is easy to profess the growth mindset belief (particularly as a math teacher), but living it can be a challenge. It means finding ways to support the student that comes in having had bad experiences that have led to a tenuous set of mathematical skills and practices. It means being vigilant about how we position students in the classroom as smart, what we mean by that smartness, and how we publicly unpack that for students to see, hear, and consider. It means speaking truth to power and pushing against fixed mindset talk and actions with other adults in the building, even (indeed, especially) when we think they do not necessarily mean harm in what they are saying. And it means welcoming students into the mathematical work by finding the assets that they bring to the classroom rather than identifying what we perceive as deficits.

Perspective 3: Mathematics Itself Is a Social Construction

The final important perspective is that mathematics is a social construction. This may seem like a starkly different grain size as compared to the other two perspectives shared here, but it is plays a critical role in true student-centered mathematics teaching and learning.

There are generally two poles in describing the nature of mathematics: that mathematics is a human, social construction; or that mathematics is an inherent set of natural structures waiting to be discovered. Compelling arguments can be made for both perspectives, but each also comes with consequences. If we presume that mathematics is some fully developed system that we as humans are still working to uncover, it takes away our agency as those who can create mathematics. It is simply a matter of discovering what has been laid out for us and hidden by ... whom? A higher power? Some fundamental underlying mechanic of the universe? Both?

When we position mathematics as a human endeavor, we invite and can recruit students into the work of creating that mathematics, not just doing what was left for us to do by others. This is particularly important at the high school level—students need to see the mathematics that they do and learn as useful in answering important questions in their world. They also need to see themselves as capable of creating new mathematics to answer questions that do not yet of answers, or that may not even have been thought of yet. This positioning of mathematics has implications for the tasks we choose to engage students in in the classroom, the tools that we do (and do not) make available for their use, and how we talk about the origins of mathematical ideas.

CHAPTER 2

SHIFTING TIDES

National Initiatives, State Policies, and District Demographics

A span of 30 years brings significant changes to the educational landscape. Often times, changes at the national, state, or district level can cause seismic shifts in classroom teaching. An example of this took place in the state of Maryland in the late 1990s. I [Mike] was teaching eighth grade at the time in a small rural school district with an excellent instructional team. We worked collaboratively well and we had district personnel who were well connected to the national mathematics education scene. I vividly remember a meeting hosted by our district mathematics specialist in her home in 1998 where we read and commented on a draft of a new document from NCTM, which turned out to be the first draft of Principles and Standards for School Mathematics (NCTM, 2000). Our combined 7–12 school building was the smaller secondary unit in the district (a larger and separate middle and high school were in the county's seat) and worked well to stay ahead of the curve with our instruction and creating sensible alignments with state reporting and assessment requirements.

A cornerstone of this work surrounded the Maryland State Performance Assessment Program, or MSPAP assessment. The MSPAP was a week-long performance assessment that cut across the core subject areas. For 5 days in April, our students spent three hours of their morning conducting science experiments; analyzing data; and reading, writing, peer editing, and

A Quiet Revolution:
One District's Story of Radical Curricular Change in High School Mathematics, pp. 17–42
Copyright © 2018 by Information Age Publishing

revising across content areas. The assessment was complex and demanded a multifaceted approach to preparation for our students—there was, in essence, no teaching to the test to be done. Instead, we designed regular performance assessments across content areas in the course of our regular instruction, supported students in building strong interactive and teamwork skills that would be tested during the assessment, and provided them with assessment feedback using multifaceted rubrics (including having them score their own work). As teachers, we took professional development days when the assessment was released to schools to work the assessment ourselves, to understand how the equipment worked and what the test was going to demand of our students, and to prepare ourselves to support our students as they worked hard to show what they knew.

Perhaps MSPAP's biggest challenge was not in the assessment itself, but in the reporting. The S in the acronym is a modifier of the first P—it was a *school* performance assessment system. Results were designed to be reflective of the extent to which a school, at the benchmark grades of 3, 5, and 8, was supporting good teaching and learning in general. Individual student results were not provided, which proved frustrating for some parents who wished to know how their particular student or students were performing relative to their peers across the state (not just in a single classroom). In spite of this challenge, the MSPAP was an extraordinary tool for the state and for districts. Surveys of teachers and principals suggested that it was a challenging assessment, but one that was effective in driving meaningful and significant changes to teaching and learning, curriculum, and teacher professional development (Koretz, Mitchell, Barron, & Keith 1996). The assessment was both performance-based, requiring deep reasoning and communication, and psychometrically valid, a relatively rare combination (Yen & Ferrara 1997). External studies showed increases in reading and writing activities across the curriculum, improvements in the implementation of reform-oriented instruction, and effects both at tested grade levels and those whose students were not tested (Parke, Lane, & Stone, 2002).

The MSPAP was driving meaningful change in the state and in my own school building, and yet it all changed in an instant with the reauthorization of the Elementary and Secondary Education Act (ESEA) in 2001, most commonly known as No Child Left Behind. The requirements imposed by the law for testing at each grade level 3–8 and the need for student scores brought an abrupt end to MSPAP in favor of a new system (the Maryland State Assessment) that featured multiple-choice and short constructed response items and no integration across content areas. These changes cascaded down from the federal to the state level, and subsequently to districts that quickly abandoned performance assessment efforts no longer aligned with the assessment initiative. The reform push in Maryland did not immediately evaporate everywhere, but for districts who were concerned about losing funding with poor performance, progressive instruction was dealt a death blow.

This chapter serves as a prelude to describing the Quiet Revolution in Holt by describing the ways in which the national, state, and local landscape changed over time. This chapter provides a critical backdrop for understanding the relatively stable forward momentum of Holt's high school mathematics program over time. The contrasting case of Maryland is an important one to bear in mind. Across thirty years of changes at national, state, and district levels, what factors supported Holt in continuing to move forward? In what ways did Holt interpret and respond to broader initiatives in ways that preserved their core values? As you read, consider the ways in which your district has responded to these same or similar initiatives. How did those actions preserve the educational priorities of your district's mathematics program? What tensions and challenges were present that were difficult or impossible to balance? How might your district respond to future changes in ways that allowed a curricular revolution to move forward?

We organize this chapter by four major eras, each of which has a start and end marked by major policy landmarks in mathematics education and education in general. As you read, keep notes in your Radicalization Handbook about how your district or another that you may have been in responded to similar challenges at the time.

BEFORE THE PROFESSIONAL DEVELOPMENT SCHOOL PARTNERSHIP: PRE-1989

The national landscape. The Professional Development School (PDS) partnership with Holt High School provides us with a convenient and important marker at which to examine the Quiet Revolution's humble beginnings. The year 1989 is also an important landmark in the national context of mathematics education, as it marked the first audible shot in what was intended to be a national revolution in the teaching and learning of mathematics. Understanding the run-up to the publication of the National Council of Teachers of Mathematics (NCTM) Curriculum and Evaluation Standards for School Mathematics (NCTM, 1989) and the demographics of Michigan and Holt serves as a context for the work outlined in subsequent chapters.

At the national level, the 1980s represented a time of renewed direction for the National Council of Teachers of Mathematics as a driver of mathematics education policy, and a sustained effort to identify coherent principles and guidelines for a K–12 mathematics curriculum. Publications such as An Agenda for Action (NCTM, 1980), A Nation At Risk (National

Commission on Excellence in Education, 1983), and The Underachieving Curriculum (McKnight et al., 1987) framed the issue as a matter of global competitiveness and a pushback against the Back to Basics movement of the late 1970s, which held little to no empirical grounding as an effective teaching and learning approach. As groups from NCTM came together to identify what sort of national set of guidelines could be created, they identified curriculum as a starting point.

The work of determining what should be taught, when, to whom, and how was informed by the emergence of cognitive science research that was building a strong case that a constructivist approach to teaching and learning led to stronger student learning outcomes (National Research Council & Mathematics Learning Study Committee, 2001). In particular, the role of misconceptions and representations in how people learn suggested that pedagogical strategies that supported the learning of mathematics might be meaningfully different from strategies that supported learning in other content areas. Moreover, the knowledge that teachers needed to teach effectively might be tightly connected to mathematics content, which Shulman (1986) defined as pedagogical content knowledge. This thinking was a significant shift from the process-product research of the 1970s that sought to identify content-general teacher actions that in turn supported student behaviors and outcomes connected to learning. The transition from a behaviorist approach to teaching to a cognitive and ultimately socio-cultural/situated cognition approach was well underway.

Michigan, at this time, had a stature in these national movements that exceeded its standing in the nation with respect to size and population. Leaders in curriculum development such as Glenda Lappan and Elizabeth Phillips at Michigan State University were deeply engaged in these national conversations. Education policymakers such as Philip Cusick and Judith Lanier were asking important questions about the preparation and support of teachers, and educational researchers such as Deborah Ball and Magdalene Lampert were spearheading specific work in understanding issues of mathematics teaching and learning. Michigan was seen as a leader in mathematics teaching and learning in part because of its statewide curriculum adoption process at the time, as noted by NCTM's Marilyn Hala during the work on the 1989 standards: "So at a state level, you would naturally go to California, Florida, Michigan, and Connecticut. There were several states that you knew were bellwether states" (McLeod, Stake, Schappelle, Mellissinos, & Gierl, 1996, p. 40). The state at the time required 2 years of mathematics for graduation, including algebra and geometry content, which was at or above the standards of many states in that era.

The local situation. In the late 70s, Holt was a largely rural community of farms on the outskirts of Lansing. By the 1980s, it had began a transformation to much more suburban neighborhood outside of Lansing. A glance at the 1985 Holt High School (HHS) yearbook shows 262 seniors, almost all of whom were White. The high school was 10–12th grade, with a 8–9 junior high across town. Prior to 1989, the math offerings and pedagogy at HHS would look similar to most other high schools. This marker is particularly important, as demographically and curricularly prior to the start of the Professional Development School (PDS) partnership, Holt was remarkably unremarkable.

PDS PARTNERSHIP HEYDAY: 1989–2001

The national landscape. NCTM's efforts to guide national policy and practice around school mathematics curriculum did not start a revolution. It started a war.

The decade of the 1990s was a highly controversial time in mathematics education. The bookend releases of the NCTM's (1989) Curriculum and Evaluation Standards and the National Research Council's Adding it Up (National Research Council & Mathematics Learning Study Committee, 2001) presented an ambitious agenda and rationale for change in how mathematics is taught and learned. Curriculum development work funded largely by the National Science Foundation introduced a significant portion of the mathematics teaching corps to progressive, student-centered instruction that focused on conceptual ideas in mathematics first, followed by procedural fluency. This approach sparked significant backlash from a traditionalist sector that feared this reform movement reduced school mathematics to superficial ideas that would undermine serious, more classical mathematical pursuits (see Schoenfeld 2004 for a summary). These Math Wars, centered most squarely on curriculum and standards, opened up a number of broader conversations about school mathematics and its role in society. To what extent should school mathematics serve the interests of those intending to major in mathematics or related fields, as compared to students not preparing for mathematics-intensive careers? Is mathematical literacy a civil right (Moses & Cobb, 2001), and if so, how do we provide access to mathematical learning to all students? What is the mathematics that all students should learn?

While NCTM's standards had some influence on curricular decisions, states and districts were still doing the work of setting their own standards and experimenting with different mathematical requirements. Analyses of state standards showed massive differences in what mathematics was being taught at which grade levels (Reys, 2006; Schmidt, Houang, & Cogan, 2002). Michigan was an epicenter for the battles over curriculum and standards, in particular because two major reform-oriented mathematics curricula (Connected Mathematics Program at the middle school and

Core-Plus Mathematics at the high school) were authored at Michigan universities. Additionally, two MSU-affiliated faculty members served as NCTM Presidents during this era (Gail Burrill, 1996–1998, and Glenda Lappan, 1998–2000), making them and Michigan by association a public face for mathematics education reform efforts.

During this same time frame, educational research continued to underscore the importance of changing mathematics education in the United States. Large-scale research demonstrated the impact of rich, high cognitive demand mathematical tasks on student learning (Stein & Lane 1996). At the same time, international comparisons of classroom practice showed that United States classrooms lagged behind other nations in the rigor of the mathematical tasks that were used in observed classrooms and the quality of questions that were asked (Stigler & Hiebert, 1999). Research on teacher development began to identify ways to better implement research results like these in classroom practice, focusing on school-university partnerships and teacher preparation programs that were grounded (both figuratively and literally) in the work of K–12 classroom teachers.

The Math Wars began to wind down with the publication of NCTM's revised standards, Principles and Standards for School Mathematics (PSSM), in 2000. Seen by some as a compromise document, PSSM tempered some of the more controversial statements by the initial set of standards and clarified a number of positions in ways that brought many of the traditional mathematics advocates back into the fold. At a district level, PSSM enhanced the staying power of a number of the reform curricular programs that had been fighting for their lives during the math wars.

From a demographic perspective, a major shift took place in Michigan in 1994 with the institution of the Schools of Choice policy. Spurred in part by national conversations about increasing equity for students, Schools of Choice opened up a limited number of seats in suburban and rural classrooms for students to move from schools perceived as struggling to schools with stronger track records of state achievement as measured by the state assessment.

The local situation. At HHS, 1989 also marked the watershed moment where Perry Lanier pitched the idea of partnering with MSU to be a Professional Development School (PDS). Earlier in the decade, a national group of teacher educators created the Holmes Group to examine and reform teacher preparatory programs. MSU was a critical voice in this program; in fact, for a while it was chaired by Judith Lanier at MSU. HHS would partner with MSU to study educational issues and problems, reflect on practice, and implement new methods, programs, and pedagogy to drive school improvement. This work initiated some of the first radical experiments in teaching and learning math. For some of this history, we draw in this section on the recollections of Mike Lehman, a now-retired Holt colleague who was integral to the work at the time:

> One of the first things we talked about was our Practical Math class (the students called it "practically math") and our Pre-Algebra class. With the urging of the people from MSU (Perry and then Dan Chazan) we eliminated these classes and to our amazement our failure rates did not change much. Now that I look back at it I'm not sure why I was so surprised by this.

As the 1990s moved on, the roots of eliminating low tracks were taking hold. The link to MSU also laid the groundwork for questioning the dependence on a textbook:

> The move away from textbooks to more activity-based and group-based instruction was another of the first big moves. [Through the PDS partnership] I had a grad student (Pam Geiss) in my room each day for one year [and] I taught one section with the traditional textbook and with my lesson plans from the previous year. [I taught] another section using groups and more open-ended problems. Pam was taking notes for both hours on how the students and I reacted, the questions I asked, and holding me true to the research we were doing. It was one of the hardest and most exhausting years of my teaching but it also showed me how things could be. I now know it was Perry's way of introducing me to cognitive [dissonance] in my own teaching.

This era saw an infusion of student teaching interns from MSU (in conjunction with the development of a fifth-year full time student teaching internship), teacher-led PD work as a building, the instigation of late-start Wednesdays for protected professional learning time, an embracing of special education inclusion and coteaching, and explorations of functions-based approaches to Algebra. When I [Craig] arrived as an intern in 1998–1999, there was a clear core of nontraditional teachers in the department and a building culture of inquiry. The department conversations would often mirror the larger Math Wars conflict when beliefs and commitments about how math should best be taught would flare up. I felt insanely lucky to be able to learn from this core group, and even more so the following year when I was hired to be part of the department. Due to the connection with MSU and a supportive building principal (and later, superintendent) Tom Davis, HHS through the 90s was featured in professional literature and many teachers saw their jobs as both classroom teachers and professionals in a larger context. It was during this time that the US News and World Report featured HHS in a national article about innovative ways of teacher development (Toch, 1993) and the sitting Vice President Al Gore visited the school during his 2000 Presidential campaign.

The suburban population growth of mid-Michigan through this decade, particularly in places like Holt, taxed the physical resources of the district. Our 10–12 building was filled to capacity (probably over) with about 1,250 kids in a building that used to have 750 a decade before. I was one of many teachers when I was hired without a classroom of my own, traveling to classrooms of teachers during their planning period on a cart. The hallways were literally shoulder to shoulder, wall to wall, with students lurching along slowly between classes as one huge jostling mass. I would estimate the minority population of HHS when I began teaching at the turn of the century to still be 10% or less. Families' socioeconomic backgrounds, however, were highly diverse. New housing neighborhoods were springing up that catered to professionals working in Lansing and General Motors executives in some sections of town, while other parts of the district contained inexpensive modular home communities and low-income housing complexes. Due to what at the time was a progressive stance on special education and inclusion, we drew a significant percentage of students with individualized education plans (IEPs) to HHS from neighboring districts.

As anyone in the education field for some time knows, the year 2001 would mark a seismic shift in federal education policy. Just prior to this moment at HHS, we can see a district that is working hard to engage in meaningful mathematics curriculum reform that provides opportunities for all students. Faculty are engaged with one another in talking about teaching and learning and finding new ways to improve on their practice. It would be easy to imagine that a hard press for more measures of teacher and student accountability could have derailed, or even ended, this work at Holt. In many places whose stories will never be told, this was the case. But at the risk of spoiling the next several pages, Holt will face these challenges to come as opportunities to further their work.

THE DAWN OF NO CHILD LEFT BEHIND: 2001–2010

The national landscape. The 2001 reauthorization of the Elementary and Secondary Education Act marked a major national shift in education policy. Whereas the Department of Education was only elevated to cabinet status during the Carter administration, two decades later President George W. Bush levied the department's influence to make sweeping changes to how students and schools were assessed. This push towards more fine-grained accountability resulted in increased testing burdens for schools—whereas most states were administering standardized assessments in grades 3, 5, 8, and 10, No Child Left Behind mandated yearly testing in grades 3-8 and increased testing in high schools. Most every state was faced with reworking their standardized assessment system on a short timescale, and as such the result was a regression to more closed-ended assessments often produced by mass-market evaluation companies. At the high school level, a single measure at grade 10 gave way to trends such as end-of-course assessments that served to homogenize the mathematics content taught in the high school sequence.

Despite the moves on the part of NCTM to find common ground in the struggle between conceptual and procedural mathematics, the push for accountability via No Child Left Behind and the subsequent changes to state testing regimens led to a de facto retrenchment of procedural mathematics instruction as a broad national trend. Advancements in interactive educational technology led districts to select quick fixes in the form of individualized computer-delivered software packages that strengthened students' skills but did not attend to conceptual understanding. At the same time, studies began to emerge connected to the NSF curriculum development initiatives that consistently showed those student-centered resources caused no damage to students' procedural skills, while providing moderate to significant advantages to students with respect to conceptual understanding (Senk & Thompson, 2003). While the first- and second-generation of student-centered curricula saw strong adoption in this time frame, student-centered resources at the high school were not widely adopted, and those that were tended to last three years or less before being replaced by more traditional curricula (St. John, Fuller, Houghton, Tambe, & Evans, 2005).

Focus also began to shift towards the role of the teacher in supporting student learning in the classroom. No Child Left Behind implemented a teacher accountability regimen that focused on student outcomes, most often conceptualized as performance on standardized assessments. Two associated conditions created a situation that likely negatively impacted instructional practice focused on conceptual understanding. First, the accountability systems were state-based, meaning that each state could create and administer their own teacher evaluation system linked to their own state assessment (Stecher, Vernez, & Steinberg 2010). Second, the changes made by states to large-scale assessments emphasized procedural knowledge (Lane, 2004), leading to a teacher assessment system that favored procedural outcomes in mathematics. Suzanne Lane, in a 2004 presidential address to the National Council on Measurement in Education, emphasized the need for accountability systems to incorporate formative assessment measures to better represent classroom-based assessment practices, and for more coherent and rigorous content standards that could help drive instructional change.

Sadly, this call went largely unheeded. The end of this decade marked the rise of Value Added Modeling, which took assessment data as inputs and sought to control for student background and prior achievement to quantify the role of the teacher in improving student outcomes. Such models serve to black-box classroom instruction and de-emphasize the role of observations of teaching practice, assume the random assignment of students to classrooms, and do not account for contextual conditions such as tutoring and family support. As such, they have received significant criticism as a very narrow conception of teaching practice (O'Neil, 2016; Strauss, 2016) and an unreliable measure (American Statistical Association, 2014).

This policy-related turn towards large-scale quantitative teacher evaluation stands in stark contrast to the focus of the education research community during the 2000s. Building on stronger understandings of cognitive science and how disciplinary experts think about their content (National Research Council, 2000), mathematics education research focused on the unique nature of the mathematical knowledge that teachers use when planning for and enacting good mathematics instruction with students. Mathematics education researchers identified the construct of *mathematical knowledge for teaching* (Ball, Thames, & Phelps 2008) as a means to describe a knowledge base for teaching. This knowledge base included both the mathematics itself that teachers need to know and the multifaceted and nuanced understandings of how mathematics can be approached in multiple ways that support student learning. This line of research, along with policy recommendations such as The Mathematical Education of Teachers (Conference Board of Mathematics Sciences, 2001, 2012), began to better describe and quantify the specialized nature of teacher knowledge that supports stronger student learning outcomes.

While a consensus set of new national standards would not arrive until the end of the decade, there were local and national efforts focused on strengthening outcomes for all students related to high school mathematics. Fueled by the work of Achieve to describe college and career-ready outcomes for students, Michigan was among the states that increased graduation requirements in mathematics. The state in 2007 mandated 4 years of mathematics including Algebra II for all students. While this policy was met with significant pushback and was eventually softened (taking Algebra II over two years counted as two credits), it did represent a significant step up for most high schools in Michigan. At the national level, NCTM was working to promote meaningful mathematical activities in the high school that go beyond symbol manipulation, as represented by their policy document Focus in High School Mathematics (NCTM, 2009). Sadly, this document appeared to get little uptake outside of schools and districts that were already philosophically aligned with a problem-centered, student-centered approach to mathematics.

The local situation. In 2001 and 2002, HHS began preparing to build a new high school on land across from the 8–9 building. It became operational for the 2003–2004 school year, with the old high school being redone as the Junior High to house seventh and eighth grades. The current 8–9 building would now serve just the freshmen, and HHS would now be considered a 9–12 high school with the separate freshman campus across the street. Besides logistical adjustments, this change in retrospect had other implications. Although exciting and awesome to get larger classrooms, new technology, and space for all students and teachers, it also meant that teachers displaced on their prep were not all together in a work-space or the lounge talking together. No more travelling teachers meant we were not in each other's classrooms every day. Having teachers active in the planning, design and creation of the new building was innovative and empowering, but it also was a disruption to our thinking about teaching and learning. Longtime math faculty member Sean Carmody noted:

> I think building this building as wonderful as it is was a huge change in momentum, in terms of what our focus went from to what it became. Like the fact that it was really important as a staff to have input on what the building would look like. because one of our things at the time was the staff has input in the things that are a part of our high school, because the high school is us. But it changed what we did as a department and as a building for two years.

And while having one's own classroom would be seen by almost every teacher as an asset, if not a imperative, the necessity of sharing rooms had evolved into a shared culture-building opportunity. Sean again:

> [A]s much as I hated traveling, I actually think traveling was very helpful in terms of integrating people. I know that going into Dave [Hildebrandt]'s room and trying to do Love & Logic (Fay & Funk 1995) as a first year teacher and being really crappy at it, but having somebody there who's like no, you talked to that person really good, you did this ... it wasn't abnormal to have somebody be in the room or to be in the room while somebody's teaching, I think is a big thing.... [I]t's really natural when you're in the room with somebody else who's a master teacher, like with Dave I can go that really stunk, what can I do different you know like how could I make this thing to better because that did not go well at all, and I know you've taught that class before, that was actually really a helpful thing that I got to do for a couple of years. [Y]ou had a chance to teach in everybody's room and you had a chance for everyone to watch you, and you get different levels of feedback than when you're an intern I think when you have a colleague, and you realize that that's kind of how colleagues talk to each other and what they do in their work. But it's a little different than just telling that story. It's not storytelling, it's working on something and talking on something to get better at it in a really specific way in five minutes before you go to your next room.

Restructuring as a full 9–12 school with two campuses offered new opportunities, but it also meant combining two disparate math departments together without a common history or belief system. In the years following the opening of the new building, we had to work hard to find new ways to continue the professional culture and momentum. (We expand greatly on this transition in Chapter 3.) And, having a shiny new building brought in people to the community and having room to spare also meant we could open our doors under the School of Choice law to students outside of the Holt school district. In the years following the opening of the new high school, HHS grew to about 2,000 students in grades 9–12, including a marked improvement in diversity.

During this era, districts in Michigan handled teacher evaluations via their local bargaining process. Contracts were ratified between teachers and the Board that included the ways in which teachers would be evaluated, how often, the timeline, and protocol for a teacher who receives an unsatisfactory evaluation. Holt's teacher bargaining unit and the administrators acting on behalf of the Board worked together to craft an evaluation model that was, by definition, fair and reasonable to both sides. It comprised about 40 pages as an appendix to the Master Agreement and included a three-tiered system (one for probationary teachers to be done annually, one for tenured teachers to be done every 3 years, and one for tenured teachers who have been identified as in need of improvement via prior evaluations) where teachers were evaluated across six domains (professional knowledge, planning and preparation, quality instruction, continuous assessment, classroom environment, and professionalism). The process included an initial meeting between the teacher and administrator, a self-analysis, goal-setting, a planning conference, cocreating a plan, formal observations, a midyear conference, and a summative conference at year-end. The document outlined in the contract included deadlines, forms, guidelines, suggestions for goals, professional resources, and agreed upon procedures (e.g., for Track II, at least two classroom observations must occur at least 60 days apart). The six domains laid out specific standards and levels of performance. The intent of Track III was, as stated, to "provide a good-faith effort to support and guide the teacher to meet the expectations set for in the Holt Public School's Standards for Professional Practice for Teachers" (Ingham Clinton Education Association, 2008). In other words, both sides believed that prior to being let go as substandard, tenured teachers deserved an opportunity to get feedback, seek out support, and have an opportunity to improve.

It was during this era that Dan Chazan organized the writing project that would lead to the publishing of *Embracing Reason*. It is the evolution of the work we have done in our department since then, in the context of this changing landscape, that has led to transformations in math teaching and learning described in the following chapters. We include as a post-script to this chapter some issues with the accountability movement that was spawned in this time period and its impact on secondary mathematics classrooms.

2010 TO PRESENT: COMMON CORE, PRIVATIZATION, AND LOCAL CONTROL

The national landscape. At the close of the decade, a seismic shift once again changed the landscape of mathematics education in the country. Following on their work with college and career readiness, Achieve supported the National Governors Association and the Council of Chief State School Officers (2010) in creating the Common Core State Standards, which specified for the first time grade-level expectations in grades K–8 guided by research on how students learn mathematics. The standards were heavily incentivized for state adoption through the Obama administration's Race to the Top program, which required states to have college and career-ready standards that met Achieve's definition in order to compete for additional federal funding.

These standards met with some pushback and opposition—more frequently connected to curricular concerns, mirroring the NCTM pushback in the mid-1990s—but in general, they represented a significant step forward in bringing coherence to the mathematics taught at the K–8 grade levels. The high school standards in the Common Core State Standards for Mathematics (CCSSM) were less clear—they emphasized the need to integrate strands of algebra, functions, geometry, number, and statistics, yet stopped short of direct and useful guidance in terms of how to organize high school mathematics for meaningful learning.

The perception that Common Core represented national standards created a backlash in some parts of the country that strengthened district actions related to local control. Even as states adopted CCSSM, most districts retained the right to supersede state mandates and adopt their own standards. (The catch, of course, is the state still held the power to determine large-scale assessment content, so the risk of moving too far away from the state's standards was significant.) A by-product of the refocusing on local control led school boards, parents, and communities to focus more closely on the curriculum and teaching at their schools.

At the same time, attacks on teacher professionalism began in earnest, centered on the Midwest United States. Legislation such as Wisconsin's Act 10 (2011) weakened teacher unions and provided districts with unprecedented abilities to hire and fire teachers often without cause. The dismantling of traditional salary scales also caused teacher turnover, churn, and bounty-hunting that changed the character and shared beliefs of many districts (some for the better, others not so much). The rise in voucher programs and privatization efforts further fragmented the teaching corps, as certified teachers increasingly found themselves working alongside, and often competing for resources with, charter schools seeking to distinguish themselves in the marketplace in a variety of ways. STEM education and mathematics was a focus of a number of these schools. To better facilitate the staffing of these schools, states also began to loosen teacher licensure leading to a more diverse, and arguably less prepared, workforce overall.

The local situation. Starting in 2011, the Republican legislature and governor in Michigan moved quickly on their own version of antiteacher state laws: public school employees working without a contract were capped in salary and benefits by what the last contract had in effect; lowered the legal threshold for a tenured teacher to be fired from "reasonable and just cause" to just not "arbitrary or capricious" and upholding that probationary teachers can be dismissed at any time; eliminating tenure as a criteria in displacement, layoff, and first-recalled situations. Public Acts (PA) 102 and 175 (Michigan Enrolled House Bill No. 4627, 2011; Michigan Senate Bill 0103, 2011) prohibited teacher unions from bargaining about teacher evaluations, teacher placement, policies about recalls or hires, and discipline and discharge of teachers. Instead the law established state requirements related to accountability and personnel, such as annual evaluation for all teachers, dismissal for ineffective ratings after three years, administrator evaluations, and mandated the use and amount of student growth data from assessments in the evaluation of teachers. Other laws froze salary steps, imposed health care cost increases on employees if a contract expires, eliminated the cap on charter schools, and created Emergency Managers who could take over a "failing" school with the power to change working conditions and void contracts. PA 152 (Michigan Publicly Funded Health Care Contribution Act, 2011) capped the amount a district would be able to contribute for employee health care, shifting significant costs to teachers who had previously bargained for health premium contributions over salary. This year also saw them cut public education by $1 billion and give a 1.8 billion dollar tax cut to businesses.

In 2012, PA 53 (Michigan Enrolled House Bill No. 4929) prohibited school districts from using payroll deduction as a means to collect dues or service fees for unions. After a voter repeal, they replaced the Emergency Manager law and made it anti-referendum-proof. They raised the cap on cyber charter schools. PA 300 (Michigan Public School Employees Retirement System Reform, 2012) required teachers choose to increase the retirement health contribution out of their paychecks or cut the multiplier as a fix to a prior law that was found unconstitutional. Finally, in a highly-publicized move in the lame-duck session of 2012, PA 349 enacted the so-called "Right-to-Work" law, prohibiting unions from requiring employees to pay their fair share of the costs associated with collective bargaining and contract maintenance (Michigan Right-to-Work Reform, 2012).

These efforts continued beyond 2012, although to a somewhat lesser extent after reaching what turned out to be long-held legislative goals (Kroll, 2014). Without going into the minutiae of these and subsequent laws, suffice it to say that the political environment was toxic for public school teachers and their ability to band together to stand up for students. Deliberate attempts to under-fund public schools by shifting available money out of K–12 designated funds or tax breaks for the wealthy saw massive budget crunches in districts across the state that continued through at least 2016 (Berkshire, 2017). As recently as the time of this writing (mid-2017), the still-Republican majority is attacking the pensions for Michigan's teachers. Predictably, from 2011–2014, federal data showed a huge drop in teachers entering the workforce in Michigan. For example, enrollment in the Teacher Ed program at MSU dropped 35% during that time. Other Universities saw more dramatic drops—the University of Michigan saw a 45% decline in those years, Western Michigan saw a 67% decline, and Grand Valley State University saw 62%. These declines are mirrored in other states across the Midwest, the nation's breadbasket for producing teachers.

Specific to high school mathematics, the state backed off a bit on their 2007 stance that mathematics should be required for four years to graduate, including Algebra I, Geometry, and Algebra II. After significant pushback by some districts claiming that Algebra II could not be expected for all students, they at first allowed schools to offer it over two years, and later carved out exceptions for requiring it altogether. Michigan also widened their definition of what it means for taking a fourth year of mathematics to taking a "math-related" course, which they allowed school districts to determine. This strategy has been mirrored across the region—for example, Wisconsin currently allows a computer science course or a career and technical education course to count as a third and final mathematics credit, at the discretion of local districts.

In all, the last decade (2007–2017) has seen the Great Recession, which hit blue-collar Michigan especially hard, combined with politically motivated budget shortfalls for public schools. Michigan went from a set of state content expectations in mathematics to adopting the Common Core, which unlike many other states, did not require a huge adjustment. Among the anti-teacher politics, as strong in Rick Snyder's Michigan as in Scott Walker's Wisconsin, John Kasich's Ohio, Mike Pence's Indiana or Sam Brownback's Kansas, the state legislature flirted with a strong stance about career and college readiness with an unprecedented mathematics criteria for earning a high school diploma. Holt High School moved beyond the growing pains of its new building, but continued to struggle with how to best implement the new state requirements across all content areas as it moved from semesters to trimesters, and then back again. After the postconstruction surge of move-ins and school of choice students, the population of HHS settled in around 1,800 students as of this writing. The following chapter will look to delve more deeply into the changes and adjustments made for the HHS mathematics department during this time in an effort to describe the moving pieces that were occurring that led, somewhat serendipitously, to the place where we ended up where every student is required to pass the equivalent of a pre-calculus course to earn a diploma from HHS.

Holt Public Schools served 5,712 students K–12 in 2015–2016. Of these, 720 are students receiving special education accommodations (13%), 2,142 qualified for free or reduced lunch (38%), and 50 were homeless. White students make up 65% of the population. Of the district's students, 1,154 are students that live in Lansing and come to Holt under Michigan's "School-of-Choice" Law. About 200 others come from other surrounding districts, making about 25% of the students in Holt nonresidents. The community has five elementary schools, two 5-6 buildings, a 7-8 building, and a 9-12 high school. It is with this backdrop that we pick up the story of how we have thought about serving our students in Holt and how that work has evolved and changed over time.

Table 2.1.
Holt High School 2015–2016 Poverty and Special Education Demographics

Grade	# Students	% Free/Reduced	% Spec Ed
9th	482	35.1	11.0
10th	453	33.8	11.7
11th	450	30.0	11.6
12th	471	28.0	14.2
Total	1,856	31.7%	12.1%

YOUR TURN: RADICAL CONVERSATIONS

The overhead music in the grocery store is not not something that you think about unless it is particularly good or particularly jarring, but it can have profound influences on your mood and your shopping habits. State and national policy is the overhead music of teaching—it likely has a larger effect on our day-to-day teaching than we realize. Taking some time to have retrospective conversations about things like graduation requirements, state policy changes, assessment shifts, and national policy initiatives can be a productive means to unpack how those changes have caused perturbations in local classroom practice. Conversations about policy can often focus on negative changes. Your task is to facilitate a retrospective conversation that notes both positives and negatives and drills down to how those changes influenced classroom practice, even in very subtle ways.

To start, bring your colleagues together and work on creating a timeline —start as far back as your most experienced faculty member can remember. This could include time when faculty members were in other districts, regions, or states. On that timeline, list changes that took place, and discuss what the nature of the change was and how it influenced practice. This can also include local policy changes, like shifts in grouping procedures, the introduction of new courses, or the addition of faculty members that effected change in the department and character of the school. It is likely that younger faculty members may have experienced some of these changes as students themselves—invite them into the conversation to talk about what they remembered about these changes as students and how they might have influenced their learning.

Bring the conversation to a close by zooming out and talking across the events on your timeline and how they influenced practice. Ask your colleagues to consider what's the same and what's different about how each of the changes influenced classroom teaching and the decisions of the math faculty and school. Discuss what the faculty would like to do the next time a change is announced—what sorts of discussions and actions might they engage in to prepare proactively for a change, to work with administration at the local and state level around the change, and to push back in principled ways that draw from previous experiences?

INVESTIGATION AND REFLECTION ACTIVITY 2

After engaging your colleagues in conversation, identify at least three key points on the timeline where major changes took place, or key changes that you know or expect are on the horizon. For each of those items, create scenarios in Table 2.2. In the "What if?" column, identify an aspect of the key point that might change in nature. In your reflective writing, spin out what effects the possible changes in constraints would have on the mathematics teaching and learning in the district. In the "What if not?" column, spin out what the effects of removing this policy (in whole or in part) might have on mathematics teaching and learning in the district. We've provided you with some examples below.

Table 2.2.
Investigation and Reflection Activity 2: What If/What If Not?

Key Timeline Issue or Policy	What If...?	What If (Not)...?
Our state end-of-course testing, last revised in 2011, is very procedural and not well aligned with our standards	*... we de-emphasized performance on the test with students?* Would they be anxious about the assessment? Would they be familiar enough with the item types? We might be able to expose students to a variety of question types while still doing the mathematics in the ways that we want to do it, and make it just a regular event rather than a focus on the big, scary test.	*... federal policy changes to erase the testing mandate?* End-of-course tests do allow us to get a consistent, if limited, view of what students know across sections and teachers. Could we institute a smaller-scale test with a wider variety of answer types? What would we put on this test? How often would we change it? Would it be individual, small group, both, something else? Who would mediate disputes between teachers about content?
Large influx of students from neighboring districts in 2013 due to state policy changes; concerns about these students' background knowledge	*... we shifted from a deficit to an asset mindset with these students, and truly embraced them as a part of our community?* The lamenting of this change sits atop a desire to get back to the kids we used to have, which were more homogeneous and because of that, it was easier to look past individual differences and treat them more as a collective. What if we were to put a stronger focus on understanding where our kids are and meeting them there, while maintaining high expectations?	*... this never had happened, and our district looked now much like it did 5 years ago?* One outcome of the shift is that it has brought forward the fact that we do not necessarily serve all students well and tend to teach to the middle. It also exposed some underlying differences in some of our teachers' perspectives on how to support diverse learners and what such students may or may not deserve/be entitled to. We all say that all students can learn, but do we all live that in our classrooms? I do not think we would have opened this question up if not for the change in our student population.

Postscript: On Accountability

My [Craig's] career was spawned concurrent with the so-called "account-ability movement" in education. When I started teaching in 1999, the notion that a state would have standards for all of its schools to follow was new, as was the fear that having similar standards for all kids meant local communities would no longer decide what was best for themselves. Initially, the move to standardized experiences was a practical (transient student populations within the state would experience less disruptions) and moral argument (shouldn't students in any place have the same educational opportunities?) with which it was hard to disagree. By the end of my first decade in education, standards were a given (to the point where national ones were soon coming) and the focus shifted to policing attentiveness to these standards, and later generalized to assessing the school program's effectiveness and in the most recent years, to individual teacher impact.

It would be irresponsible to make the case, as in other professions, that all teachers and educators are great at their jobs. Anecdotal evidence can be problematic when coming from the mouths of teenagers or disaffected parents without training in education and pedagogy, but we know there must exist some teachers who shun all evidence about best practice, do not put in the time necessary to do a great job, or have long lost their passion for working with students. And when your son or daughter has only one school experience, you cannot blame parents for wanting to be hypervigilant about weeding out Bad Teachers, even when they are often mythical caricatures from popular culture more than reality. Particularly when they are fed this narrative from lawmakers with a desire to make public schools appear to be "failing" so reformer friends and business partners can create for-profit "alternatives," it can be an overwhelming perception.

Reality, however, is as always much more complex. Is a teacher who works hard and wants students to learn in a way that is different from the way(s) you or I would approach the subject matter considered bad? Is it malpractice to have a certain philosophical belief about the way work should be evaluated and grades assigned? What does it even mean to do the job of educating students in mathematics well? And how will we be able to tell if/when we see it? There is, of course, a well-defined certification and evaluation process in place for teachers already that is charged with ensuring initial quality and continual monitoring. Is that process failing? Insufficient? Unions are often blamed as a barrier to eliminating poor teachers, but in every case I know of teachers do not want subpar actors sullying their profession. In truth, the breakdown tends to be with adminis-tration being unable to find evidence (Is seeing one lesson fall flat enough? Do student complaints indicate subpar or high quality? How much time is fair to show improvement?) or being qualified to identify poor practice in secondary subject-specific content areas. Absent of criminal abuse or neglect, it's really hard to identify which teachers would better serve society in a different profession.

The confluence of complexity, political motivation, and societal hysteria came to a head in NCLB, where the federal government under George W. Bush decided that students simply needed to be tested to see if they were learning enough under each of their teachers. Ultimate accountability, and a naively clean solution to those outside of the profession. When legitimate concerns were raised from those on the inside with knowledge and experience about evaluation and assessment, they were dismissed as teachers simply wanting to avoid scrutiny. When data came out showing that even with "value-added measures," (wherein a teacher's value was determined relative to where his or her students were at when they began) ranking teachers through test results was essentially no different than randomization (Baker et al., 2010), politicians persisted. In Michigan, the Republican-led legislature made using standardized test scores as a significant part of a mandatory yearly evaluation for all teachers, eliminated prior tenure laws, and made local collective bargaining of evaluation models prohibited.

And so, there was this weird time when the rest of the country was catching up and accepting the tenets of best practices in secondary math ed that we at HHS had been experimenting with for years, and yet much of our time and effort became finding ways to keep providing these experiences against an increasing administrative pressure to impact the kinds of data that show up on the state-determined, NCLB-mandated 11th grade test (which was, at this time, the ACT and, most recently, changed to the SAT). The superintendent at the time created immense pressure to get better test scores (which showed up on state-created district comparative websites) since public consumption of them as evaluations of the school meant better advertising, and thus more students (with their state foundation allowance).

We made arguments that simply comparing average scores on the mathematics portions of the ACT among local districts with varied sizes, financial support, and community socioeconomic status was problematic: that just having access to the mean rather than the median or measures of spread was an important part of the comparative story; that the test has bias toward procedural and rote skills over conceptual understanding; that the test is timed. And to some degree, these appeals to logic worked in as much that we were still given the space to teach the way we, as professionals, knew we ought to. But instead of unwavering support, we felt a tenuous allowance.

Here's why this is a critical consideration. We saw our students, many if not most of them from poor, minority, or undernurtured educational backgrounds, engaging in real mathematics in complex ways on a near daily basis. Dissecting strategies. Arguing about ideas. Making sense. Reasoning. We have no "low" tracks hiding students. We have real assessments that require kids to write, explain, and prove they truly understand the concepts to be able to earn credit. (And to be fair, we also have a few students in each class that we have not yet reached.) And they understand this math in ways that no program I know of in the area can even touch. Through all of this, however, we were barely above state average on the ACT-based school accountability assessment. Neighboring districts that teach in rote, procedural ways killed us on that measure. And then some administrators and parents took that as evidence that we are doing the wrong things. And then they see the things we do that look different from these neighbors and from their remembrances of mathematics in the 1950s, and all of a sudden teaching mathematics really well becomes a courageous act of rebellion again. The accountability movement, when measured the way it has been, is biased toward poor teaching practices, and this a Big Deal. Here are the reasons why, framed as critical issues:

Issue #1. Tests that can really measure understanding of real underlying mathematical concepts and practices and tests that can be done on a large-scale to compare schools are (as of yet) mutually exclusive sets. I have little faith in the ability of technology to truly change that – I often have to talk to a kid one-on-one to be able to discern how much they really understand and how much they are parroting snippets of memories from class. But if it is developed and becomes reliable, the costs will be prohibitive. So at some point, we have to recognize that good assessments cannot exist on a scale that would allow us to compare widely.

Issue #2. Poor assessments do not just not measure true understanding well, they actually honor poor understandings and misrepresent them as superior. Consider the following parable:

> Two children are riding in the back seat of the family car, zoned out after hours on the road. Their zombie stares at the screens in their lap are interrupted by a voice from the front seat, eagerly pointing out a rainbow on the horizon. One of the sons, after a moment of consideration, proudly recites "ROYGBIV" and goes back to his binary world. The other son asks his mom if one could reach out and feel a rainbow. The mom responds, and as her son's questions continue, her description gets more detailed. Over the next hour, the two engage in a conversation exploring light waves, atmospheric refraction, and prisms. From now on, when this son sees a rainbow, he sees so much more. Now imagine a scenario after time passes, where researchers want to know to what degree kids understand natural phenomena, and the question they ask is, "What color is next to red on a rainbow?"

The first son will no doubt get this correct. The second son may reason that the colors in a rainbow are ordered by wavelength, and since there is only one correct answer, red must be one extreme. Since he has heard of "infrared" and "ultraviolet," he reasons that they may be together on that end and selects "violet." Researchers will conclude that ZombieSon is the one who Gets It and the other needs to have his mother replaced with a State Appointed Emergency Manager.

Issue #3. Students that understand complexity is often unintentionally but consistently punished on these assessments. Consider the following exemplar that came to mind working cross-content with some teachers of Economics.

> One key concept in the course is the relationship between the marginal number of products for the quantity of labor, or marginal cost for quantity of product. Consistently, students have a difficult time comprehending the graphs of these relationships, and the Econ teachers were asking us for our thoughts on why that may be the case. Immediately, it became clear that the teachers were moving between the graph total production or total costs and talking about the rates of change of these graphs- something the majority of students have not yet had a chance to really consider much in their math courses.

So, like in the Rainbow Parable above, one fix could be to just tell students that the marginal cost graph looks like a Nike swoosh, come up with a memory trick like "Nike is marginal," and get almost all of the students to get that correct when assessed, but know virtually nothing. Alternatively, you could take several days, and create experiences for kids to consider building graphs of the actual rate instead of the total so they comprehend the idea of how marginal production or cost is built off of the total production or cost and be able to sleep at night. However, you have then used up class time that means some other ideas are not developed as deeply or gotten to at all ("opportunity cost" in the parlance of our Econ friends). Additionally, a student being asked on an assessment to identify the graph of the marginal cost would need to take several minutes to think through the ideas and build the graph out of that understanding, taking precious time that a student with no understanding beyond Find The Swoosh does not need. And finally, a student that understands and builds the graph has a probability of making an error, and thereby could select an incorrect answer (often a purposefully placed distractor).

There is no doubt that if the goal is to come out looking good on accountability timed tests, there is notable momentum to have your students robotically chant *ROYGBIV* and be sure not to muddle their thinking with confounding things like connections, reasoning, higher-order thinking, depth, or substance. Particularly when the tests are timed, there is no room for considering, comparing options, recreating ideas, or thinking. Robots win in this world. And teachers who avoid understanding and instead show Tricks get high-fives. And "highly effective" evaluations. And lay-off protection. And merit pay.

Issue #4. Like it or not, for whatever reason, tests like these (even the best ones with quality-control and bias-avoidance measures) correlate pretty strongly with socioeconomic status. This has been generally shown over and over again in a myriad of ways, and it is true also with the scores in the Intermediate School District for which Holt belongs: for the 2015-16 school year, the correlation coefficient between median household income and percent proficient on the SAT mathematics test was .88 (mischooldata. org, 2017).

I cannot escape the logical conclusion from this then: that somehow, these tests instead measure innate qualities or things that correlate with high SES, like speed, cleverness, reading ability, or confidence. And to some degree, this is by design—when originally authored, the ACT and SAT was meant to predict who would do well in a postsecondary environment. You know, like those that are quick, clever, read well, and confident. And even more so, to sort out those that would from those that most likely would not. There is little question as to why, then, the math portion of that test is filled with questions that ask for symbolic manipulations and clever substitutions instead of general understandings of function characteristics or behaviors. Or, like the ones below, you could completely understand the intended content but miss the question under time constraints (ACT, 2015, pp. 30–31).

· The square below is divided into 3 rows of equal area. In the top row, the region labeled A has the same area as the region labeled B. In the middle row, the 3 regions have equal areas. In the bottom row, the 4 regions have equal areas. What fraction of the square's area is in a region labeled A ?

F. $\frac{1}{9}$

G. $\frac{3}{9}$

H. $\frac{6}{9}$

J. $\frac{13}{12}$

K. $\frac{13}{36}$

· A dog eats 7 cans of food in 3 days. At this rate, how many cans of food does the dog eat in $3 + d$ days?

F. $\frac{7}{3} + d$

G. $\frac{7}{3} + \frac{d}{3}$

H. $\frac{7}{3} + \frac{7}{3d}$

J. $7 + \frac{d}{3}$

K. $7 + \frac{7d}{3}$

If $x{:}y = 5{:}2$ and $y{:}z = 3{:}2$, what is the ratio of $x{:}z$?

A.   3:1
B.   3:5
C.   5:3
D.   8:4
E.   15:4

Figure 2.1.   Sample ACT mathematics questions.

Issue #5. When accountability is defined as outperforming colleagues, the central mode of improvement in our craft is destroyed. It is a common assumption that competition drives higher quality experiences in the world of business, and many people lack the experience to understand why that ethos does not simply apply to all endeavors, like education. But if my professional livelihood is on the line, why would I share my great lessons, high quality tasks, or neat innovations with my colleagues? The notion of having "industry secrets" in education is the antithesis of the altruistic goal of the public education system: more students learning more things.

Steve Leinwand of the American Institutes for Research and the lead author of Principles to Actions, speaking at a conference in Michigan I attended, said that "isolation is the death knell of our profession." Richard DuFour (2002, p. 14) called his such goal-oriented "collaborative teams" were "the primary engine of our school improvement efforts." Anne C. Lewis (2002, p. 488) found that "[s]chools with strong professional learning communities were four times more likely to be improving academically than schools with weaker professional communities … [w]e can no longer afford to be innocent of the fact that collaboration improves performance." Collaboration is probably the greatest vehicle for improvement that exists in our schools, and accountability the way it is defined politically cannot coexist. That would be a good question to check to see if kids understand the concept of irony, except a multiple-choice question probably cannot quite capture it.

How then do public schools ensure quality? First, we have to reconsider the underlying presumptive basis for the conversation: that teachers inherently have no reason to try and be the best they can be if left without oversight. That the impetus to be lazy, to do the least amount to collect a paycheck, is more overpowering than a self-motivated educator can withstand. But it is not just that teachers like this exist (for I do not doubt they must); it is that the problem is *so* pandemic that there is no other choice but to create an external system built around this notion. And that such a system is *so* worth having, it is better to maintain it even though it actually restraints good educational practice for the teachers that do not need to be monitored. And on top of all of that, that these "poor" teachers would be *so* motivated by avoiding punitive measures that they would now decide to work really hard and teach "well."

Given the low pay relative to the amount of education and certification required to teach, this is clearly already a flawed narrative on the whole. We have to recognize that the vast majority of educators are true professionals who know that their job is a service to society; that while hoping and fighting for fair compensation, they are in it for the good of the community. That means we ought to consider the Bad Teacher as an educational Bigfoot—an extreme and isolated case worthy of spending considerable time and energy to locate, but rarer than many people imagine and nearly impossible to know for sure if you've seen it. And while it is imperative that we locate and do something with these TINOs (Teachers in Name Only), it is equally imperative that we do not interfere with the true quality work of the vast number.

If we believe (as we do) that quality education comes through collaboration, experience, and continual learning about the craft, then our educational accountability needs to honor these attributes. We need to replicate decades of collective bargaining that created contracts that honored and incentivized expertise, experience, and continued education. We also need ways to deal with the exceptions to the rule so that these TINOs can be identified fairly, given opportunities to improve, and then dismissed if necessary. It can no longer be the case where underperforming teachers get reassigned or given a gentle letter of recommendation to find a job in a neighboring community; it has to be professionally acceptable in education to council a TINO out of the profession when kids' educational experiences are on the line.

For educators in the field of mathematics, it is time to step up. The time is now for the notion of "underperforming" to include poor math ed teaching practice—a nice, funny, responsible hard-working teacher that disseminates procedural knowledge to students via lecture notes and rote practice can no longer be absolved of his or her professional sins. There is enough evidence and widespread acceptance in the larger math ed community to include pedagogical approach as a qualifying characteristic for teacher evaluations.

And, when certain standards are spelled out and expected, then meeting these standards is not dependent on your students outperforming other teachers' students. Competition is no longer the name of the game, opening the door for the single greatest way to create and sustain top educators—collaboration. Hiring practices and evaluations based on content-specific pedagogical best practices can get the right people in the door, and collaborating for student success can push them to continually improve. Teacher contracts can incentivize experience and further expertise. And when a critical number of teachers in the department or building (or district or state) are existing at the level knowledgeable experts recognize as effective, then two other phenomena occur. The first is that incredible peer pressure will now exist once that critical mass is reached. When I first got to Holt High School this was an unspoken yet palpable experience, and one that taught me what is expected to do this complicated job well. Being accountable to colleagues I respected very much has much more impact than manufactured concern about how my kids do on a state test that conflicts with our teaching goals. The second is that the staff will see themselves as professionals, and this self-replicating atmosphere will drive and sustain growth as colleagues begin presenting at state and national conferences, authoring articles about their thinking or experiences, and researching teaching practices. Administrators will learn that treating staff as professionals and experts creates more positive momentum than coercing improvement through bringing in speakers or programs to "give" professional development. Teacher contracts can even financially encourage teachers to become leaders in their field through impetuses to present, write, or become involved with teacher training programs.

CHAPTER 3

MAKING IT WORK

The Implementation of Ongoing Curricular Reform

This chapter details the ways in which changes to the mathematics curriculum at Holt were translated into courses; how student supports evolved and changed as the student population changed; and how these changes were disseminated, supported, and at times challenged by school administration and the community. This chapter is the essence of the quiet part of A Quiet Revolution, examining ways in which teachers made these changes work by focusing on student support and reasonable, achievable change.

The work of assembling a curriculum, translating that curriculum into courses, and populating those courses with students has a very particular look and feel in the majority of the nation's high schools. Curricular decisions often center on textbook selection, a high-intensity event that usually only happens two to three times over the course of a career. Teachers gather, review samples, hear about important bells and whistles as described by a publisher representative, consider broad alignment between the contents of the text and standards, debate relationships between the text and testing requirements, and make a choice. Districts might invest in some initial professional development related to the text to support initial implementation, but this work typically is short in duration. Translating the curriculum into courses is often an automatic move, as the United States high school landscape is still dominated by the Algebra I-Geometry-Algebra

A Quiet Revolution:
One District's Story of Radical Curricular Change in High School Mathematics, pp. 43–65
Copyright © 2018 by Information Age Publishing

II sequence of courses, followed by a path to Calculus and a rotating cast of adjacent courses for students who do not enter the calculus sequence. These courses typically range from more rigorous options like Statistics to survey-style rehashes of middle and high school content of topics that seem to have been most troublesome for students (from a teacher's perspective). Courses are divided into levels, so that the Honors track can help identify the students most likely to take on the black diamond trails at the top of Mount Calculus, with others taking the blue square or green circle paths through their mathematics education—never too high up on the mountain, never too steep a descent. Populating students into those courses and tracks means setting thresholds for adequate performance to move to the next step, usually determined by a course grade. Failure to meet that threshold means you move from the black diamond down to the blue square, or the blue square to the green circle. Or perhaps the student stays in the ski lodge for a year.

As you might imagine, Holt's curricular process is markedly different, and we believe there is much for districts to learn from their example. This process extends beyond simple matters of curriculum and touches on teacher hiring decisions, the professional learning community, and the ways in which the faculty develop and maintain shared values and vision. This chapter and the one that follows will provide you with insights into Holt's process and history, first with a chronological narrative in Chapter 3 and next in Chapter 4 with the voices of longtime Holt mathematics faculty that further describes the ways in which conversations and decisions evolved.

Our intention is for these two chapters to serve as an illustrative example to outline the sorts of conversations and initiatives in which districts can engage to make meaningful student-centered change. For example, one of Holt's current supports essentially takes a population of habitual failing students and supports them so that their learning outcomes look the same as the general population the second time through. We will describe the work Holt has done to destigmatize retaking courses, moving away from models that make a "dummy" or slower version of the courses and segregate kids. Holt's perspective is that if they were going to have a student take two years to learn the material (and there is no shame in that) they want them to be in the general population and learning the course like everyone else. We describe Holt's practice of not always issuing failing grades, but granting a "not yet" credit of elective math as if these students simply surveyed the course the first time. This results in courses with 75–80% of the students learning and passing a rigorous course, and the remainder either students who did not do the hard work necessary to learn and earned their E and will retake, or students who worked but through no fault of their own (maybe the background was insufficient, maybe they have learning disabilities, etc.) were not able to demonstrate enough understanding for credit, and will get a "not yet" and re-enroll. The idea behind sharing this story is not so that you can plug the same system into your own district. Rather, it is an accounting of the challenge and the current solution (that starts and ends with meeting student needs) to help you think about how you can begin conversation that lead to similar solutions that best fit your district. We start with a review of the curricular changes and broaden out to consider the ecosystem within which those curricular matters live.

HOW DID WE GET HERE? AN OVERVIEW OF HOLT HIGH SCHOOL'S CURRICULAR HISTORY

Tasked with summarizing the confluence of events that conspired to create the circumstances under which Holt High School (HHS) has moved forward since *Embracing Reason*, I [Craig] imagined telling a relatively linear story—this happened, and we chose to do this, and as a result we tried this, and then this occurred, and so on. As it turns out, seldom do retrospectives present themselves so cleanly, and our recollections of any sequential order are foggy or incomplete; at the time, none of the HHS math faculty were thinking of recounting a history. Similarly, which event influenced what decision or vice versa is difficult to parse out years later. In reality, few events of educational change happen at a particular time; some evolve over a period of time informed by other forces; some occur in concert with other changes, their synchronicity allowing either to exist; and some happen in response to potential occurrences that never came to be. Instead, what we have is an intersection of various forces that, over a long and continuing period of time, molded the day-by-day operations of our department. Even with a fractured and nebulous plot line, it is the dynamic responses of a collective group of math teachers to these forces, trying to stay true to a shared vision of what it means to learn mathematics, that emerges as the protagonist.

In the prior chapter we tried to trace out some of the landscape in which this volume of work exists. We examined an evolving context on a national, state, and local scale. But to really understand, or rather to *feel*, how this played out from within a particular location, it seems helpful to imagine all the variables at play. First, we have our "math department," which year to year varied in size and personalities as student teaching interns came and went, members retired or were reassigned to different buildings or took leaves, new teachers were hired, and teachers from other departments were assigned spare sections of mathematics. A fixed core stayed constant, but conversations about direction and policy were always influenced by a slightly different cast of characters, each with his or her own changing understanding of how to be most effective teaching mathematics.

Another variable is the people to whom we were responsible—the building principals, curriculum directors, superintendents, and so forth. Much of the district curricular change comes through their offices, and we spent a considerable amount of time helping people in these positions understand the basis for the work we do. And those people changed over (sometimes more than once) in this time period, at different moments, and the degree to which we had to fight for what we believe in varied thusly. Even the way the same people in those same positions felt about the work we do varied considerably over time, as they are also responding to administrative mandates, messages from the intermediate school district (ISD), external pressures, and their own deep-seated beliefs about what learning math looks like. The other variables are best grouped in large chunks as simply *external* and *internal*. By external, I mean just the things that came from without that we had to respond to, such as changes to State law, Board policy, or structure of the school day. By internal, I mean the topics we noticed ourselves about what we thought needed attention: failure rates, how students respond to writing prompts, lacks of perseverance or motivation, and so forth.

In contrast to this undulating, overlapping, multidimensional pseudo-Venn diagram of factors, there exists a handful of constants that our work centers on, and has ever since the early empowered days of being a Professional Development School (PDS) school. First is being self-reflective and willing to share vulnerabilities with colleagues in a continual effort to improve. Second is our deep commitment to staying true to a particular pedagogical stance for mathematics education. And third, an insistence that expertise from the ground up is the most effective way to improve education. (We explore these constants in more depth later on in the chapter.) These moorings are central to the continual instigations and reactions to stimuli that maintained a forward, focused direction. In short, there are lots of moving parts as there always are, and between constant conversations about how to teach well and maintain our vision through a changing topography, we constantly strived to do what was right for the students as informed by our commitments to teaching mathematics really well. And in that process, through incremental responses and generating our own internal momentum, we ended up at a place where every student earning a degree from HHS had to demonstrate pretty deep understanding of content in a course equivalent to Precalculus.

For the narrative of the curricular evolution to take shape, I am not going to try and chronicle all of the minutiae, but rather give enough of the supporting detail for readers to see some of the relationships between all of the variables involved in such a dynamic system. There is a fine line between enough detail and extraneous exposition, but the hope is that understanding the environment provides a setting to the story that grounds the plot in reality. Like any good educational *noir*, the setting itself is often a character in the story.

The PDS transition and its effects on curriculum. The story begins with a quick flashback to the PDS heyday in HHS of the 1990s. The existence of state standards was years away still, and a few in the math department were experimenting with teaching differently. A core group began thinking about how to shift to a discussion-based pedagogy, and to insist on reasoning and problem solving as key components to courses. As part of a building-wide initiative initiated through PDS work for writing across the curriculum, courses started including explanations, write-ups, and reports as pieces of assessment. The building had a progressive ideology of full inclusion for students with disabilities. The department started becoming active in the larger math education community. The department had University of Chicago School Mathematics Project (UCSMP) textbooks, but some started re-thinking the utility of having a textbook that gives students all the answers at their fingertips. Marty Schnepp tinkered with sequencing and the development of concepts in his Calculus class. The eighth-ninth grade Junior High was tracked at that time, and a group of students arrived to the high school having been in *Transitions Math* for 2 years. For them, Sandy Callis and Dan Chazan cotaught an Algebra course Dan created about linear and quadratic relationships that is function-based and oriented around student questions and experiences (whose story is documented extensively in Chazan, 2000). They began to reconceptualize "curriculum" as all the experiences students have as part of working together, including their prior experiences with math and perspectives on learning math. They housed course documents on a shared drive accessible to the department. Over the last few years of the decade, other teachers shared in the development and teaching of the Algebra project initiated by Dan and Sandy.

I walked into these efforts as a new hire in 1999–2000, teaching the Algebra project. I have vivid recollections of printing out a unit on the shared drive, taking a pile of papers to my apartment to spread out on the floor of the spare bedroom, and trying to organize them in a way that made sense. The image is reminiscent of the relieved-of-duty cop obsessed with the mystery and having the walls covered with case files, suspects, and clues; except instead of a serial killer I was simply trying to find a coherent storyline through a logical progression of ideas. These Algebra classes, filled with students who had been told they were not good at math and had life experiences I could barely comprehend, did not always look pretty. But I watched these kids slowly, and suspiciously, open up a bit to a math class that did not feel like what math had been to them so far. It was during this time that I learned to value collaboration and conversation with colleagues about what we saw students do, as my inexperience meant I was often surprised by the ideas students offered in class and needed to talk out ways to plan subsequent experiences to build on these ideas.

Successes of these early efforts pushed conversations about expanding nontraditional approaches and challenging assumptions about learning. We wondered whether strategies that were good accommodations for students with disabilities might be good for all students. Could all students benefit from being pushed to think differently about learning mathematics in this way? It was decided that the Algebra project course could be used in eighth grade for the "on-track" students as well, and this was our first experience trying to share curriculum and pedagogy with teachers that had not been a part of the development and conversations surrounding the goals of a function-based algebra program. What we learned was that in the absence of the shared vision and access to the professional development work we had been doing, the project when handed over was quickly routinized and made more traditional—vocabulary sheets were created, lessons were added to "teach" what they believed students needed to know prior to doing the assignments built to generate that knowledge, and the web of ideas and supporting tasks became a pre-printed sequenced set of handouts that were followed regardless of what questions students had. It was an early indication that reform-minded teaching could not be coerced with reform-based curricular materials; that in teaching as with learning, telling does not work. For the teachers that bought in, they were able to replicate many of the experiences for all students; but even then, without the collegial support structure, philosophical commitment, and experiences about what it looks like when playing out in a classroom, it was difficult to sustain the program in the junior high building.

Expanding beyond the Algebra work. In parallel, we also began to spread the tenets of the Algebra 1 project beyond just the population of students who came to us that did not have access to algebra in eighth grade. When I began teaching the Algebra 2 sections with like-minded colleagues, we began to adapt that course to be more consistent with this Algebra 1 approach. Some teachers, however, stayed with their more traditional approach and used the UCSMP Algebra texts. For several years during this period, we had broad agreement over what content needs to be covered in the courses, but sections varied wildly based on who was teaching and their personal beliefs about how students learn best.

Over the next few years, grades 6–8 eliminated low tracks, and went to a program called MathScape to replace what had become of the Algebra project course there. In our building, we adjusted to teach a course that integrated topics of Algebra and Geometry (IAG) for the transition years where students had a fragmented sequence, as the new curriculum was not phased in. After a few years, we were able to fade that out, as all students could come to our building ready to take Algebra 2 (with some careful work at the beginning to "functionalize" their understanding of linear relationships and build up their understanding of quadratics). During this era, we also were given funded sections to try and support struggling students. There were dedicated sections to help students that may have been behind, but that did not create a slowed-down or lower version of the course. Instead, in the first year, I had a section of an Algebra Support Class (ASC) that was comprised of students from throughout my other classes who would come to me in their last period and spend more time on the content together. Besides having a second hour of Algebra to work and think in their day, we also incorporated lessons on how to have successful group work, problem solving, writing effective explanations, staying organized, and so forth. Over the next few years, this ASC course structurally changed as scheduling students correctly was prioritized below the amount of work it took to create said schedules, and the course devolved to having kids from any teacher who may or may not be working on the same idea or topic at the same time. Thus, instead of more time working together with their teacher on an idea, the course became a bunch of students working individually on whatever their ASC teacher happened to be doing. The next year, it became even more egregious, as all kids that may need extra help were thrown into one or two sections together regardless of what math course they were taking. (This approach resembles many of the more typical extra-time intervention strategies that do not tend to see strong student learning outcomes.) At one point, math faculty member Brian Vessell had kids from the IAG class, Algebra 2, and the junior-level Precalc (FST) class together, meaning it could be little more than just a math help room. Nevertheless, the faculty persisted in thinking creatively about how to use the funds the district had earmarked for supporting math instruction in the high school. One year we used the dollars freed up from eliminating the corrupted ASC sections to have some math sections cotaught with two math teachers. This served those classes of students well, but the real benefit was having newer hires experiencing what it was like to teach nontraditionally in real time with someone who had been doing for a while.

A physical restructuring provokes a conceptual restructuring. While these curricular changes were going on, plans were starting to take shape to address the overcrowding the 10–12 building was experiencing. Community growth and an increase in school-of-choice students (many drawn by the full inclusion and support services we offered for students with disabilities, and others from sharing a large district border with Lansing Public Schools which was getting, sometimes unfairly, a negative perception) had gotten us to beyond capacity. This situation was so acute that Marty Schnepp taught his sections in an empty classroom in the elementary school next door, and those of us in their first few years rotated with carts to rooms during people's planning periods. The Holt community narrowly passed a millage that would fund a new high school in a field across from the existing feeder building, an eighth and ninth grade junior high. This building would be split and would now house just the ninth grade. The eighth grade would pair with the seventh grade in the old high school building, creating a new junior high. This change meant that the ninth grade Geometry teachers, who had been existing as their own entity across town, would now be joining the high school math department. We saw this as an opportunity to revisit our curricular sequence (would it not be cool to instead do a ramped-up Geometry in 11th grade *after* the algebra sequence?) and had many tough conversations during this time. There was a wide variety of perspectives about math education represented, and the influx of experienced teachers that had not been a part of the professional work we had been doing for so long made the conversations at times contentious. We took pride in our ability to have passionate disagreements and still remain friends, but our local analog to the Math Wars would at times challenge that feeling.

The new high school and building configurations opened for the 2003–2004 school year, with the redefinition of HHS as a 9–12 high school across two campuses. We did decide as a 9–12 department to continue to have Geometry in 9th grade, so in practice that meant that the ninth grade Geometry teachers usually met separately during department time and we continued to operate as distinct entities. During the next few years, there were some other fundamental changes of note. The last vestiges of textbook-based courses faded. An extremely supportive, knowledgeable, and pragmatic Superintendent retired, and the new Superintendent had a different idea about leadership that was more top-down and focused on large-scale assessment scores as a proxy for the learning our district was providing. The Board required that all classes give common exams, to ensure consistency and rigor across courses. As part of that, our building principal, unable to avoid the pressure from the Superintendent, added a requirement that all exams at the high school have ACT-like multiple choice sections in the hopes that experience with questions like this would translate to higher scores on the state-mandated test. The result for us in math was that we could no longer have questions on the final that were specific to the learning experience of that particular class. For example, we would no longer have questions like on fifth hour's final that asked, "Explain why Mike's strategy for finding the 'rate of the rate' did not work for all of the quadratic tables we examined." A more critical loss was that the oral exams, which had already been a teacher-to-teacher difference in the Functions, Statistics and Trigonometry (FST) and some of us had expanded from the old Algebra project into their sections of Algebra 2. They could no longer exist unless they could be done for all sections across the board, and logistically we could not have judges and space to run performance exams for a thousand students. So, the Honors sections of the FST course kept them for a time, and the rest of us began thinking hard about how to have an exam that still honored explanations and measured true understanding, and then also had the highest quality multiple-choice section we could muster.

Another challenge in this time was that the district, worried about operating an additional building, looked carefully at its budget. The massive influx of school-of-choice students after opening the building helped immensely, but the district administration saw some opportunities to curb long-term financial outlay. For teachers, the most critical decision was to eliminate coteaching by special educators, and instead adopt a "teacher consultant" model (where, by law, special education caseloads could be much larger). Up to this point in secondary grades, teachers in Holt's special education department would spend four of the six periods in a day assigned to a class that had a significant need for support. Over the years, I worked with many great coteachers that not only made the course better for the students, but helped me become a better teacher. Now, special educators have ever-growing caseloads (via waivers where even the federal cap can be raised) and are not in classrooms as teachers, but in offices meeting with their students as a sort of specialized counselor.

In response to many of these challenges and the opportunities offered with some staff turnover, our department began thinking about ensuring consistency and reworking some of the courses. Inspired by a workshop about Wiggins and McTighe's (2005) *backwards design*, we set about creating documents that identified the Big Ideas in each unit, the Enduring Questions, related content from the state standards, and examples of the kinds of assessment tasks students would need to do by the end of the unit. We spent district-provided professional development days crafting tasks for places we saw as weak spots in our sequence. We ensured our final exams had a constructed response section and many questions like we would expect on a given course's unit test (in addition to the school-mandated multiple choice section). We used these common assessments as a sort of quality assurance, knowing that if the expectations on the final were high, individual teachers would strive to get students to be able to explain, write, justify, and demonstrate proficiency on critical ideas from the course. Our department swelled in size in response to a huge bubble of students coming to HHS, and for probably the first time, we were able to say "we believe" instead of "some of us believe" when it came to a math department ideology for the building. Educational momentum for STEM (science, technology, engineering, and math) topics and pushes about college for all led to Holt adopting a third year of mathematics as a graduation requirement.

State-imposed change: New standards and tighter funding. In 2007, the state implemented their Grade Level Content Expectations, which were an ambitious and rigorous stance about the topics required for kids in Michigan. Coupled with this, the state also defined what it took to get a high school diploma in the state of Michigan, and it included 4 years of math: Algebra for 2 years, Geometry, and another math or something math-related in the senior year. They also ramped up to 3 years of science (including Physics or Chemistry) and social studies to go along with 4 years of English Language Arts. It was a revolutionary stance by the state, intended to make Michigan's students able to attend college and be in high demand for the workforce of the new technological century. In preparation for this focus on core content, HHS staff debated going to a 5-period day trimester schedule to preserve room for elective courses. Full-year courses would become two trimesters, and semester courses would become one trimester. Our department shared our concern that the mathematics was getting expanded and requirements higher for all kids, and even though classes would go from 60 minutes a day to 73, a change to trimesters would mean less time than if we did 60 minutes a day for 180 days. Recognizing the dilemma, math was allowed to have three trimesters for the required courses in freshman through junior years.

The next several years, we were teaching courses to all kids where we had 73 minutes a day for the full year. For the first time, we saw passing rates at highs we had not seen before, with a much more diverse and varied set of students. Unfortunately, the original intent of the trimester schedule began to break down as other foundational courses in other disciplines were given the same three-trimester luxury. As more courses required for graduation went to full-year, three-trimester long classes, the schedule started to resemble what we had before, only with just five periods in a day instead of six. To compound some of the stresses we felt in an elective versus core academics zero-sum game, the recession that began in 2008 hit Michigan hard. The unemployment and stress of the families we served increased, the state budget was in chaos, and people began migrating out of the state. By 2010, the state's population had dropped 175,000 people from just 5 years before (U.S. Census Bureau, 2012). These components combined with the politics of the Republican leadership in the state meant a huge financial crunch for the state's public schools. At Holt, this meant not replacing retiring teachers, running classes above the class size maximums, and exploring cheaper alternatives like online education. This, among other factors like tough contract negotiations and top-down administrative edicts, led to labor unrest and an overall feeling of loss of trust and efficacy from teachers.

After about 5 years on the 5-period day, the district changed back to a 6-period day in the 2012–2013 school year. For us, that meant a hard conversation about how we could eliminate 18% of the class time with as little impact on the students as possible. We knew we did not just want to blow through the ideas faster, nor did we want to cut the time we spent helping students think about what good explanations look like or letting students wrestle with ideas. To accomplish the task, we probably did a little bit of all of that. Most notably, we chose to cut the unit on Power Functions in the sophomore year and instead just focus on what negative and fractional exponents mean (sacrificing examining what functions like x^m behave like for different rational m). We also decided to swap a unit that we felt was not going as well as we wanted in the beginning of the junior year (Rational Functions) with Exponentials and Logarithms from the end of the sophomore year. This made sense to do anyway, as students coming out of Polynomials could then study Rationals without having to spend as much time after a summer break revisiting characteristics of the polynomial functions that make up the numerator and denominator. It also allowed us to have Logarithms (which we always taught paired with Exponentials) in the junior year, when they could study other transcendental functions like trigonometrics. Finally, studying the characteristics of rational functions minimized the loss of studying power functions that year.

The advantage Michigan had when the Common Core State Standards (CCSS) came out was that the high level of Michigan's state standards. This meant that schools that had already integrated them had only minor content tweaks to face. The advantage we at HHS had was that the Standards for Mathematical Practice, which were now required in all math classrooms, were already largely in place. For us, what the CCSS did was legitimize decades of work already put in, and give us an opportunity to say that the discussion over what should be done and how had been settled. It felt as though we were finally at a spot to get past some of the unrest and focus again on how to get more kids to learn more stuff together. But as soon as we started getting our feet on the ground with all of these changes, the district administration—over the objection of a majority of staff and a significant portion of the community—decided to bring the freshmen from their own campus into the high school building, and put the seniors instead in the campus across the street. The hope was that it could offer neat pseudo-college opportunities and atmosphere, keeping and drawing students (and their state allowance funding) to the district. While the benefits and drawbacks could be debated, what it did for sure is create one more distraction from talking about teaching and learning as we had to consolidate who teaches the senior courses (for example, in the year prior to "the switch" I had taught AP Statistics, but had to give it up unless I wanted to go across the street for part of my day) and send two colleagues across the street while gaining the four Geometry teachers. If nothing else, it created a 9–12 department for the first time, as our department chair,

Marty Schnepp, was one of the teachers to move to what was now called the Senior Campus. It did, however, also mean we had some work to do to incorporate a handful of people that have not been involved in the same conversations.

We relied heavily on the Common Core statements to talk not about what a math classroom should look like, but rather how to improve at getting our classes to look like that. This was further strengthened with National Council of Teachers of Mathematics (NCTM) release of Principles to Actions (NCTM, 2014), both of which played a large role in helping bring new Curriculum Directors and another new Superintendent on board in recent years. The next chapter of this curricular story involves the work we are doing to reshape the Geometry course as a large group now, and the public relations we will do with a Board of Ed that has completely turned over since the "switch." Stay tuned for that.

The story of curricular change in the math department at HHS for me is bookended by my first year, when Dan Chazan came to our department and orchestrated a writing project that became *Embracing Reason*, and wrapping up writing for this volume in my 19th year of teaching. Obviously the story continues, and the cliffhanger of what comes next at HHS is really interesting to just a few. The real intrigue is in the next chapter to be written that is part of *your school's* story. *Your* prequel is undoubtedly different—although maybe with similar threads or themes—and your setting is another place and time. The Enduring Question is, *What lessons can we take from across all of our stories to inform actions that lead to a productive way forward?*

The epilogue to this sequence—this history, this context—is a quirk of happenstance that was looked at as largely an annoyance at the time. But it served to further confuse the storyline and obfuscate the path that got us from then to now, and is a good example of why some revolutions are quiet: when the state released the high level standards and upped the requirements for graduation, they created examples of course organizations of the standards. In doing so, they decided that having Algebra 1 in eighth grade for everyone can now simply be called "eighth grade math" and the first course of algebra taken in high school should be referred to as Algebra 1. Naturally, the next course would then be called Algebra 2. So when the State made the statement that Algebra 2 was required to get a diploma, they were asking all kids to take up through topics that had been called Precalculus or Functions, Statistics, and Trigonometry for all recent history. This is why over the subsequent few years, the state hedged quite a bit on requiring Algebra 2, making addendums that this course (but not any of the others required across all subjects) could be taken over two years. And then they defined Personal Curriculums as potential exceptions for students, they carved out Algebra 2 as the lone example of a course that could be omitted altogether for students with IEPs. Public misconceptions of learning math notwithstanding, these changes in names for the courses coincided with so many other changes, that we went ahead and just did it at HHS. A student in Holt would go from Geometry, to Algebra 1, to Algebra 2, to AP Calc or whatever senior year option they wanted (as a math or math-related course was still required by law). As you can imagine, we spent a lot of time having to explain the naming changes to parents and counselors who were used to decades of socially determined (yet vague) meanings of these names. Transfer students were problematic as well, since many schools did not appear to make the course name changes; so a student would transfer in to Holt taking Algebra 2 at their school—which there was the first high school course for algebra—and be enrolled in the same content class here, but titled Algebra 1. Ultimately, we renamed the four required semesters of Algebra as A, B, C, and D so as to avoid the issue with some schools having changed names and some having stayed the same.

So here is the surprise twist to the story (and maybe you are much wiser than us to have seen this coming): Even though state law is that all students need to take Algebra 2, and there were public documents at the time defining what that meant, we can find no one else that made the name change that turned Algebra 2 into Algebra 1, and Precalculus or FST into Algebra 2. Maybe because they needed to be able to keep their textbooks and the titles on the covers, or maybe because they simply never saw anything except simplified what-do-we-need-to-do filtered through their administrators that missed that detail, or maybe even schools do not really believe all kids can learn high levels of math, or maybe the emotional and logical ties to those titles are just too strong. Quite by accident—and yet, the outcome is not surprising—we ended up being the only place we know of making sure every kid got through 2 years of Algebra (semesters A–D) just like the law states, and the topics of those courses were updated versions of Algebra 2 and FST ... just like the progressive vision of the state requires. And we just did it, not realizing how revolutionary we were being, because we thought it was just the way it was going to be. Turns out, years later, it still is not the way it is outside our borders, despite the fact that this conception largely aligns with the content expectations laid out in Common Core. The innocuous lettered names seemed to defuse most freakouts from people making assumptions about ceilings for some kids that mathematics must have but other topics do not. It is funny how the same choice feels different depending on if you were part of a whole-scale adjustment or if you were doing it by yourself. But it does not change the fact that we have a decade now of Holt students that have done it, with no black hole opening up or tears in the fabric of reality.

The origin of this book was the realization that other schools were not making the adjustment we did, and in fact many of them were using the change as an opportunity to challenge the notion that all students could learn rigorous mathematics. And watching the state get pushback and hedge on whether they should really expect all kids to do "Algebra 2," I felt like we should tell our story, so citizens and policymakers could know that with intentional supports and an informed approach to teaching, all kids can be expected to learn whatever level is chosen. In truth, I would not balk if it was decided that all students need to understand Calculus (or polar coordinates, or algebraic coding, etc.) by the time they graduate high school—the kids and classes are the same, the mathematical topic they are trying to think through is just later in the sequence. Mike gets full credit for thinking about how an idea for an article making sure people saw an example of real kids being expected to learn through Precalculus topics could actually be a larger vehicle for guiding change, which you are reading now.

A STUDY IN SUPERFICIAL CHANGE: WHAT RESEARCH HAS TO SAY ABOUT CURRICULAR REFORM

Holt's story of curricular change, like the work of teaching and learning mathematics, is far from a set of procedural steps to follow that map out a clear, linear path from start to finish. Indeed, Holt's story is cyclical, iterative, adaptive, and messy. It stands in stark contrast to what we know from research on curricular implementation, particularly in high school mathematics. Curricular reforms at the high school level in most districts tends to be driven by the adoption of curricular materials as a catalyst for change. Using text adoption as a driver, however, is an effort doomed to failure based on what research shows about successful and meaningful high school reform that supports student learning.

Research investigating the processes of curricular and pedagogical change at the high school is sparse at best. The dominant focus of the mathematics education research community has been strengthening elementary mathematics teaching and learning, an important and pressing need. The research that does exist at the high school level presents three clear and powerful themes: student-centered pedagogical practices correlate to stronger student learning; students in more integrated mathematics courses outperform their peers in a more traditional course sequence; and curricular and programmatic change is fragile and fleeting. We briefly explore each of those themes, and relate and contrast those research themes to the work at Holt High School.

Student-centered pedagogy. The publication of the NCTM Curriculum and Evaluation Standards (NCTM, 1989), followed by the Professional Standards (NCTM, 1991), portrayed a clear picture of what good teaching and learning should look like. These documents (and many that have followed) emphasized authentic problem solving, inquiry, and student discourse as key components of an effective mathematics classroom. These policy initiatives were followed by sustained teacher professional development through the National Science Foundation's Strategic Initiative and Math Science Partnership grant programs. This work was further buttressed by the development of progressive mathematics curriculum materials that embodied a student-centered pedagogical approach.

The mathematics education research community studied these curricular reforms from a number of standpoints. Senk and Thompson (2003) synthesized pilot studies of the effectiveness of the five new high school curriculum materials, all of which explicitly supported student-centered pedagogy. The pilot studies on all five of these curricula showed overall positive results on student learning. In general, students did no worse or slightly better on procedural assessments of learning, and (where available) outperformed their peers on more open-ended conceptual assessments. McCaffrey and colleagues (2001) examined the performance of tenth-grade students in a district implementing two different reform-focused mathematics programs while maintaining a third programming track that represented traditional content and pedagogy. Analysis of student learning favored the classrooms in which student-centered teaching practices were being implemented, in conjunction with the two reform-focused curriculum materials. These studies, together with more recent studies of high school pedagogy (e.g., Boston & Smith, 2009), provide evidence that student-centered teaching approaches are effective in promoting stronger student learning outcomes for high school students.

Integrated mathematics: Breaking the Algebra I-Geometry-Algebra II canon. NCTM's (1989) initial set of standards broadened the focus of mathematics in grades 9–12, making strong cases for the meaningful inclusion of topics such as statistics, discrete mathematics, and mathematical modeling for all students. To address this imperative, the curriculum development efforts funded by the National Science Foundation (NSF) all made use of some form of integrated course structures rather than the typical Algebra I-Geometry-Algebra II (AGA) high school sequence. The AGA sequence is generally nonexistent in other parts of the world, including countries with which U.S. mathematics performance is either competitive or superior. This course sequence was initially designed as a filtering system, with the ultimate goal being to identify students who would be poised for success in a high school calculus course. In conjunction with the clear message that calculus is not the intended summit for high school mathematics students, organizations like NCTM, NSF, and the Mathematical Association of America pressed for the overhaul of high school mathematics course content to reflect a more integrated pathway. Subsequent policy documents like the Common Core State Standards for

Mathematics (National Governors Association Center for Best Practices, & Council of Chief State School Officers, 2010) have continued to advocate for such an approach.

The implementation of the NSF-funded reform curricula provided significant opportunities to compare student performance between integrated and traditional course structures. A suite of studies from the University of Missouri found that students in more integrated curricular sequences were more successful on multiple assessment measures (both procedural and conceptual), and that curriculum-related factors had stronger influences on student achievement than teacher-related factors (Grouws, Tarr, Chávez, Sears, Soria, & Taylan, 2013; Tarr, Grouws, Chávez, & Soria, 2013). As students progressed to the collegiate level, there were no significant differences between students with a more integrated experience as compared to a traditional course sequence in the number and difficulty of collegiate courses taken (Post et al., 2010). Taken together, these results suggest a net positive immediate impact of a more integrated approach to high school mathematics curriculum, with no identifiable risk to future mathematical advancement in college.

The fragility of curricular reform at the high school. A landmark study by St. John and colleagues (St. John, Fuller, Houghton, Tambe, & Evans 2005) provided the field with in-depth portraits of curriculum reform at the high school level. Up until this time, anecdotal evidence had suggested that mathematics education reforms at the elementary and middle school levels were much more durable than high school reforms. In examining a wide range of curricular implementations, the authors found that the average high school mathematics curriculum reform effort lasts less than three years. Reform efforts often hinged on the work on a small number of individuals (or a single person), and pressure from the outside of a district or a lack of central administration support swiftly and effectively rolled back reforms, even in districts in which the faculty were fully on board with change. St. John and colleagues (2005) noted that professional development, a shared set of pedagogical values, and a variety of aspects of administrative support were critical features in the more successful implementation sites. This work resonates with McCaffrey et al. (2001), who noted that student-centered pedagogical practices were as important as reform-oriented curriculum materials.

These three factors together suggest a number of important conclusions about high school mathematics reform. They show that while a student-centered curriculum may be a necessary condition for meaningful, sustained change, it is not a sufficient condition. Curricular and pedagogical reform must proceed hand-in-hand for meaningful and durable change to take place. These studies also show that high school reform, perhaps more so than at other levels, can be derailed by a lack of administrative support or by tenuous, limited buy-in on the part of other faculty. This finding is particularly relevant given the sentiment, often pervasive among high school mathematics faculty, that teaching methods and strategies are a matter of choice and style rather than of research, beliefs, and values. Chalking choices up to style and individual character provides teachers license to enact less effective pedagogical strategies. Combined with the perception of unimpeachable subject matter expertise among high school teachers, this often translates into choices to use alternative texts and resources that undermine a curricular/pedagogical reform that a school is trying to implement. In short, reform at the high school level begins with, and indeed hinges on, a shared set of values that are enacted every day in every teacher's mathematics classroom.

Spillane (2000) examines mathematics curricular reform from the lens of district policy, and finds many of the same results. His analysis of the relationships between state policymakers and local district implementation in multisite case study in Michigan revealed important factors that constrained implementation of a student-centered pedagogy. Spillane indicates that on the whole, there was more attention at the district level to the mathematics content than the process of teaching mathematics. This is evident in a retrospective sense in the privileging of the 1989 Curriculum and Evaluation Standards (NCTM, 1989) as compared to the Professional Standards (NCTM, 1991) and the all-but-forgotten Assessment Standards (NCTM, 1995). NCTM did attempt to bridge this gap by integrating pedagogical and assessment-related messages into Principles and Standards for School Mathematics (NCTM, 2000), but the content continued to be king. Curriculum development efforts in the mid-1990s spearheaded by NSF were compatible with a student-centered pedagogical approach, but such resources stopped short of clear, meaningful messages to teachers about how to teach effectively. Indeed, the most vocal higher-education critics in the Math Wars that pushed back against these curricular reforms centered on the mathematics content far more than the pedagogical approaches (Schoenfeld, 2004). This resistance at the national level may stand in contrast to more local experiences that focused on unfamiliar pedagogical approaches, with parents often being concerned that collaborative learning approaches were not serving some students well and allowing others to skate by. Even still, the most vocal critics such as James Milgram and David Klein, focused their attacks on specific mathematics content that they felt student-centered curricular approaches were not supporting students in learning.

In the Michigan case studies, Spillane also noted that pedagogical attention tends to be superficial at first and does not effectively challenge the canon of procedurally-based mathematics. This means that even in the face of a curriculum whose effectiveness rests on being implemented using student-centered pedagogical approaches, some teachers are likely to use old, procedural, and incompatible teaching methods. From an administrative policy standpoint, the emphasis of curricular change focused on form over function—were the instructional materials visible to administrators in classrooms when they walked through to observe. The results of this emphasis on materials over pedagogy is the application of rote pedagogical techniques on meaningful mathematical tasks, reducing them to sets of steps to be memorized and executed (Stein, Smith, Henningsen, & Silver, 2009). In such situations, visible artifacts like manipulatives can be used in procedural ways (see "The Case of Monique Butler" in Stein, Smith, Henningsen, & Silver 2009 for an outstanding example). Group work becomes the sharing of procedural steps with a focus still on correctness rather than communication, debate, and grappling with meaningful mathematics (e.g., Lotan, 2003). In such situations, students do not see the point of working together, parents complain, and teachers back off of the superficial approaches and gravitate back to lecture and seatwork.

Figure 3.1 synthesizes these findings from classroom and policy-based research to compile a typical workflow in secondary mathematics education reform. This flawed process is typically viewed to be linear. Discuss and agree on courses, standards, and curriculum materials that will undergird a program. Create structures intended to ensure homogeneity of experience across teachers and students: pacing guides to ensure "proper" use of the curriculum materials and common assessments to measure outcomes "fairly." Across these two phases, there are significant risks to failure —teachers may become demotivated and burned out, feeling a loss of freedom, claiming their kids cannot rise to the challenge of the new curriculum (and without pedagogical reform, they cannot), and remaining stagnant in their pedagogy because there is no impetus to change. The reform may end there. The implementation of this effort tends to initially function on aspects of form, such as the use of manipulatives and student learning. At this phase, the implementation may be superficial, with rote manipulative use or the devolution of meaningful mathematics tasks to be rote procedures that are not worthy of group engagement, thus undermining and ultimately devaluing cooperative work in the classroom (with classroom management sometimes suffering collateral damage). It is only if all these pitfalls are effectively cleared that a reform can move to the function-oriented aspects, truly engaging students in meaningful problem solving and communication. The critical pitfalls illustrated in the figure give a clear indication of why so many high school reform efforts that ascribe to this linear, top-down approach fail.

The process described in the Holt narrative is markedly different. From the beginnings of the curricular change with Dan and Sandy's reforms in Algebra 1, the function-focused aspects of pedagogy were at the core.

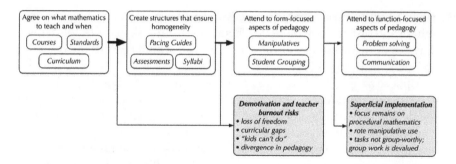

Figure 3.1. A typical workflow in secondary mathematics education reform.

An important learning issue was identified—in this case, supporting students who had spent 2 years in the Transitions program who were not being well served as they entered the high school. Rather than thinking about a structural solution or a specific tool to try to remedy that issue, they cut right to the heart of the matter by rethinking ideas about what it meant to make meaning of function-related concepts and to promote a discourse-based classroom pedagogy. The Holt story shows how that work radiated and iterated into discussions of how we might better agree on unit content, on structures for courses and students' movement through them, how that could influence the ways in which courses were scheduled and students were grouped, and how students at both ends of the prior achievement spectrum could be supported in realizing their potential. You might have noticed that the parts of the narrative in which standards and state requirements shifted, the policy change was a bit player in the story. At center stage was always the notion of how those outside changes could be leveraged as opportunities to improve the teaching and learning of mathematics. The student-focused pedagogy remained at the center of the work that the Holt faculty did do and continue to do. Figure 3.2 represents the process in which Holt has engaged. You will notice that most of the elements of the diagram are the same as Figure 3.1, with only the descriptors in boldface being slightly different. But the starting point and the cyclical and iterative nature differ greatly from the typical workflow as represented in Figure 3.1. To enact change in the sorts of ways that Holt has, it is critical to bring conversations about pedagogy and student learning to the center of your school's work.

YOUR TURN: RADICAL CONVERSATIONS

The differences between the typical curricular reform process and Holt's, as illustrated in the two figures in this chapter, are striking. Simply based on probability, it is likely that the processes that have taken place in your own school and district look more like Figure 3.1. Your task in this Radical Conversation is to bring the conversations in your school towards the center of the circle in Figure 3.2.

Choose a topic that you have been discussing in your building related to the teaching and learning of mathematics. Try to select something that's located on the outside ring in Figure 3.2, and your goal will be to frame that conversation in the context of function-focused aspects of pedagogy. For example, if you are in a curriculum adoption cycle, you might discuss the curriculum resources that you are considering. If you are grappling with issues of grouping students either in courses (tracking) or within classes (for collaborative work and conversation), you might discuss this issue.

Whatever issue you choose, start the conversation by focusing on why you, individually or collectively, are making the choices that you are making or considering. In terms of our ability to help students make meaning, why are we leaning towards curriculum resource A over curriculum resource B?

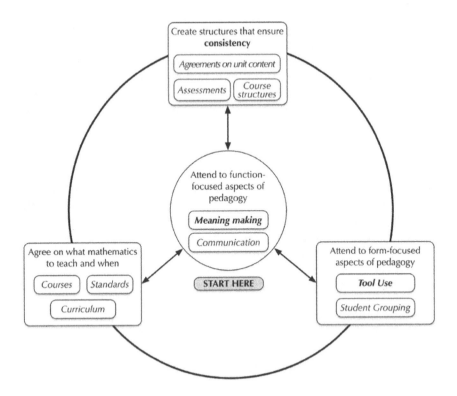

Figure 3.2.   Holt's approach to secondary mathematics education reform.

In what ways do we expect these materials to support stronger communication between our students about mathematics, and how do we know? What strategies are we using that are working well to get students talking? What strategies are not working well? Are we providing our students with equal opportunities to engage in good conversations about important math, and if not, what could we change structurally to get better? Why are we choosing to use (or not use) the math tools that we are using (or not using), and which students have access to them?

Take notes during this conversation. Identify the idea on the outside of the ring that you intended to address. Record the questions you planned in advance to steer the conversation towards the center. Write a brief reflection on the ways in which the conversation did address issues of function-focused pedagogy, and the ways in which the conversation remained on superficial concerns. The Radical Conversation in Chapter 4 will provide you with some opportunities to dive more deeply into these ideas with your colleagues.

INVESTIGATION AND REFLECTION ACTIVITY 3

Your Investigation and Reflection Activity for this chapter follows on from the Radical Conversation. But here, we'd like you to turn the lens on yourself. Choose three aspects of your teaching related to the outside ring in Figure 3.2. For each of those aspects of your teaching, identify what you are currently doing that seems to be working well for your students. Make sure to detail how you know it's working well, and for whom it is working well (Is it working for all students? Most of them? The ones that come in with strong prior knowledge? My struggling learners?). Following that analysis, identify a pedagogical strategy that you would like to try or refine that will help you address this particular idea in a better way. Remember that this is an experimental attempt—it is okay if it is not successful! Stick with the change for a while, and if you need to revisit this activity to make a new change, do so. Table 3.1 can help you to organize your thoughts, and contains an example.

Table 3.1.
Investigation and Reflection Activity 3: Reflecting On My Pedagogical Choices

Outer ring idea *(about the mathematics to teach, the resources I use to teach it, the surface-level strategies I use)*	What I am doing now that is working well, how I know, and for whom it's working	What I would like to change
In my geometry class, I use compass and straightedge methods to teach the unit on triangle congruence.	This strategy works well for the students who are more linear thinkers and who like procedures. Students are able to reproduce the steps and they know which conditions result in congruent triangles and which do not. They are generally able to justify their thinking using triangle properties. I can see this in the proof and proof-like arguments I ask them to make. They also can solve context-based triangle congruence problems even when it's not really obvious what they should use. My students who struggle with steps, or who tend to gravitate towards visual representations, seem to have difficulty with triangle congruence.	I would like to integrate transformation-based approaches to triangle congruence, and to make use of physical manipulatives and technology in addition to my existing practice with congruence. I wonder if this might help my students who get hung up on the steps and are not seeming to think about what congruence means and how we can justify it. A transformation based approach can be mathematically rigorous but it will be a very different set of arguments for my class!

(Table continues on next page)

CHAPTER 4

LEADING AND SUSTAINING RADICAL CHANGE

Voices From the Trenches

The process of change over time at Holt is clearly different from the average district, as we have described in the previous chapter. What we have illustrated are the macrolevel storylines—the general trajectory of both Holt's evolution and the ways in which most districts go about change. In this section, we zoom in to hear the voices of some of Holt's veteran teachers describing the details of change. In particular, we focus on what the process of change has looked like at Holt from moment to moment, year to year; how Holt approaches the work of staffing and onboarding new teachers in ways consistent with the department's beliefs and values; how the department handles failure and loss; and the advice that the experienced Holt teachers would give to teachers like you trying to instigate revolution within your building and district.

The excerpts below come from several conversations we conducted with current and retired Holt High School mathematics teachers. The voices you will hear in the following section are Craig Huhn, Sean Carmody, Dave Hildebrandt, Steve Lawatsch, Mike Lehman, Marty Schnepp, and Spencer Sullivan (an intern in the district and new hire in 2017). Mike Steele appears in a few excerpts as the moderator of one such reflective conversation in 2017. Transcripts of these conversations were edited only for readability – any changes to actual spoken content is noted using [brackets].

A Quiet Revolution:
One District's Story of Radical Curricular Change in High School Mathematics, pp. 67–92

ON DECISION MAKING

The first excerpt below talks about the nature of decision making at Holt. Often times, decisions about changes to courses, assessment, and even the focus of teacher professional development are made by principals or central office administrators. A little bit of top-down decision making is certainly inevitable within a district—the state might change a graduation requirement, the board might pass a budget or push an initiative that will have implications for faculty work. But the ways in which the teacher leaders at Holt approach this work is significantly different from the average district. In the excerpts below, Holt faculty discuss how they determine that there are changes that need to be made, and how that process unfolds. We pick up the conversation after what had been some historical recounting of changes that had occurred over the past two decades. Mike begins the conversation by asking the Holt faculty to draw across these cases.

Mike Steele: So drawing across those different time points those different changes, how does it go ... amongst the faculty when it's clear that there's a change that's needed somehow. How does that process, come about? How do you realize there's something that needs to be done and how does it go?

Sean Carmody: My guess is that those have not been uniform, each of those changes have been unique, it hasn't been do this, do this.

Dave Hildebrandt: Changes in external expectations, changes in personnel, changes in our student performance... The best is sometimes when we take a look and see how our students ... are performing, what they are succeeding on, what kinds of questions they're answering and how they're answering, making us rethink how we approach the mathematics ... the multiyear picture of what we want students to do and succeed at. But that gets impacted by if there are textbooks that we're using or certain personnel are using certain things or wedded to a certain approach, if they leave or if there's a change in our schedule or a change in our district policies, state policies.

Marty Schnepp: The district mandating common assessments made that be a little bit more difficult too, because we had differences from classroom to classroom that would probably be teacher based. I know when we first did our common assessments we had all of the teachers who taught the course look through the standards, look at what our units would look like, look at what it would look like in the end, we had a lot of discussion on what those common exams were gonna look like, we were coming off of a number of us using oral exams and that's a pretty big difference with what we are doing now.

Craig Huhn: it wasn't just exam stuff though, there were a couple years where we pulled out data and we sat in the Ed Center ... I remember spending a full day trying to find a higher order polynomial real life context or create one that [would] motivate the unit.

Sean Carmody: That also could have been our formative assessment group that met too.... We were trying to come up with what the standards, the "I can" statements ..., and what it would look like to really do this. The push from ... some of us was, what does it look like contextually for people to do this, [and] to apply this in some kind of way.

Marty Schnepp: I don't know if this is accurate or I'm trying to make sense of things in my mind, but it feels like ... we kind of recognize these issues, but the changes don't get made until some of these larger systemic things happen. [For example,] formative assessment stuff was brewing in the background and then the mandate to have common exams ... worked its way in. So we chose to try to make the best out of that and take those sessions and do "I can" statements and things like that. There's a third thing I had in my head that sort of fit that mold where, you know we are all incredibly active professionally, we ... all know what's going on out there.

Dave Hildebrandt: Look at our department meetings and the readings we've been doing every year, it's under the same umbrella but [we've had] a new approach, a new way of looking at things. The [growth mindset] stuff that we read ... 3–4 years ago ... we used it as a lens for how we're looking at everything. And then we shift. Those things are always in play, they are always there. But we're always adding layers to what we're doing.

What can we take away from this conversation about how Holt identifies areas for change? First, note the statements made by Marty and Dave at the end. The fact that the faculty regularly engage in reading research literature is important. Dave notes that the ideas in the literature serve as lenses for considering changes to practice, and that those lenses endure and add to one another rather than one hot new initiative replacing the previous hot new initiative. The regular engagement, consideration, and revisiting of research-based ideas suggests that these ideas will inform conversations about changes to curriculum, assessment, and course offerings.

Second, it's important to note that the evolution of change is ongoing. Marty and Sean recall that some ideas were worked on and not immediately implemented, such as the formative assessment work, only to be brought forward and revitalized at fortuitous moments, often corresponding to an external change. In many districts, when a change comes down, the faculty meet to discuss how they are going to implement that change. Frequently that conversation revolves around what they will drop or suspend within their current set of initiatives so that they can do this newly-mandated thing. Change at Holt tends to follow a different pattern. When an external event takes place, the Holt mathematics faculty ask themselves (consciously or unconsciously), what are we doing or what have we thought about in the past that might be a good conceptual fit for the change we are being asked to make? This move has two important effects: it allows the faculty to take ownership of the change, and it provides continuity by building on an idea that had already been in play at some level for the group. This is a simple shift in decision-making strategy and processing that has the potential for transformative results.

On Staffing: Recruiting New Teachers and Onboarding Them

The staff at Holt High School is, to a person, an extraordinary collection of thoughtful, reflective, and dedicated individuals. Your school and district may not look the same way. Holt's staff did not evolve by accident—there are specific steps that the mathematics faculty have taken to identify teachers who will be compatible with Holt's approach and to foster and support their integration into the professional learning community. The two excerpts that follow unpack the perspective of the Holt staff with respect to hiring and onboarding new faculty. The partnership with Michigan State University certainly plays an important role in this work, as Holt often has the opportunity to preview potential hires when they are assigned to the school as student teaching interns. Our first conversation takes place immediately after the veteran teachers noted that just about everyone on the math staff at present was trained at MSU. An intern who was hired and about to start his first year, Spencer Sullivan, was also present for this conversation.

Mike Steele:	That's a pretty impressive list, both the fact that the majority of the faculty are MSU trained and then this list of people [who were not] that have come and gone, is really interesting. How do you as a department think about looking for your Spencers, looking for the people who are going to fit, other than they came to MSU, they were an intern here? What other things do you look at, what do you think about? ... Both looking for someone at the point where we have a vacancy and we're looking to interview and hire, but also the onboarding process, what does that look like.
Sean Carmody:	I think we're getting more explicit about that? Craig's pretty good about talking about disposition ... in terms of something that he's looking for, and how you talk about how people learn math. I know we've looked at grades in terms of math classes, and math majors as opposed to minors, as being criteria we'd like to have for people to fit.
Dave Hildebrandt:	I think there's something to having people who are familiar with multiple levels of mathematics so that they can process that and make that connection with students more readily. When I'm working with interns the one thing that I want them to do, the hardest thing is for them to understand where each part fits into the whole. So when we're doing a specific lesson I understand because I've been doing this for several years not only how it relates to the unit but how it relates to the course and how it relates to the subsequent courses you know maybe 2–3 years down the line, and it's harder to

have an intern that has that perspective, who can take that big picture approach.

Sean Carmody: And that was a mistake we made a couple of times, we tried to hire a person to fit a role that they either outgrew or didn't want to do or didn't fit into all the other things that we had. *This* teacher will work really well with *these* kids. And I think that idea of a big picture approach, people who can teach everybody, is at least for me a big thing. We think about our kids that are in Algebra C/D, we have kids that will take Calculus in a year, and we have kids who never want to take math again, and both those people need to learn how to do Trig. So as a teacher being able to handle both of those extremes and making it a good experience for all of those kids is really, challenging; if we hire somebody who is really good with these honors kids or you really are good with these kids that struggle or you know, we want somebody who is a little more well rounded.

Marty Schnepp: In the interviews, we try to get them talking. We listen really hard for, are you talking about how well you really explain math, or are you talking about how the kids are responding to what you're having them do. If it's clearly a coach first and a teacher second, those people don't even get into the pile where we come bring them in. A former principal liked perky people that had that spark, that's a phrase that my wife quoted from him in a couple interviews, and we were looking for the person who is really interested in mathematics and curious about mathematics.

A number of important ideas emerge from this exchange. First, the notion of dispositions—towards the discipline of mathematics and towards mathematics teaching and learning—is an important piece of the interview process for prospective hires. Whereas some districts give their prospective hires content tests to assess their mathematical skill, Holt takes the approach that the disposition towards mathematics is as critical a criterion as knowing the mathematics itself. In addition, rather than a checklist of skills that the teacher needs to know, we see Sean and Dave mentioning the importance of understanding the mathematical continuum—where the mathematics is going and where it is coming from. Finally, Sean's point about hiring someone versatile is important. To be able to effectively work as a part of a team and accommodate the ebbs and flows of a school, teachers need to be willing to teach anything in their subject area and all types of students. Sean alludes to times in which the district sought to hire someone with a particular specialty, and that strategy did not appear to work well. Not only does this limit flexibility in moving staff around across courses and sections, but it also implicitly sends messages about ownership of particular student populations. *These* students are Mr. X's, and *those* students are Ms. Y's. *This course* belongs to Mrs. Z., so you should not ask to teach it. On a staff that believes that all students can learn, one must also believe that all teachers can teach—every course and every kid. Steve Lawatsch, a former faculty member trained at MSU, captured the sentiment recently in a social media post:

Steve Lawatsch: This was really difficult for me when I was a [preservice teacher]. It was made harder by the fact that my field placements as a senior put me in situations that seemed to reinforce my own views of teaching math as a skill set which is done by putting the correct pieces in the correct order, rather than the art that I came to appreciate during my time at Holt High School. It is extremely difficult to convince someone that their views may not produce outcomes that are as desirable as one would hope when they have no sense of a need to change them. I had to have experiences with students that showed me the value in having a different disposition before I had that "a-ha."

Finding good potential teachers in an interview setting is one thing. Supporting them to learn, grow, and thrive is quite another. In the excerpts that follow, the veteran teachers talk about the ways in which Holt has approached supporting new teachers in becoming a part of the community.

Marty Schnepp: In terms of the onboarding one of the things we have been trying to do recently is make sure the [new] person gets one prep and to the extent we can not sacrifice that person to the hardest traveling schedule. It's a weaning process but I think that's, really significant in giving them a common prep with someone who's experienced with that course. You see so many other places where the new hires get the really challenging courses, the worst rooms, and we really don't want to do that to people, because we really want people to come in and have a solid experience learning to teach.

Mike Steele: So thinking about the onboarding, how do you think about bringing someone in with respect to the nonteaching parts of the day? There's a lot that goes on here in terms of curriculum development, in terms of these different pockets (I think Sean used that word) of activity going around. How do you think about bringing someone into those pieces of work?

Marty Schnepp: I think the Wednesday mornings [late-start departmental meeting time] probably play the biggest role.... I try to make sure it's clear when people come on that we read stuff outside of here and we embed the work we're doing in the literature of math education as opposed to the textbook we're now using or whatever popular workshop series they're putting on.

Dave Hildebrandt: I think that Wednesdays are our secret weapon. They really facilitated our ability to push ourselves, not only in the department meetings but … to really get people to be thoughtful about their teaching practice and share. The conversations that we've had about common grading, about the course materials some of the backward design questions and some of the other things has really forced us to be thoughtful about what it is we're doing. One thing I always rationalize to MSU when they ask why I want an intern is that it forces me to be thoughtful about the choices that I make, but doing that in the department is even more so. I think we're good about trying to get everyone to contribute, there are a few people who were silent and on the sidelines—

Sean Carmody: Those tend to not be newer people.

Dave HIldebrandt: —and they're not necessarily newer people, they're people that have a trajectory that they're used to. We've tried to disrupt a little bit, to get you know some other people to mix things around, and I think one thing I appreciate about this department is we're not wedded to who teaches what to a large extent.... We're really flexible and every 2–3 years we look and we make changes, so sometimes it helps to disrupt when we get little pockets of people who are same teacher doing something, but when we get someone else in they'll say, well explain it to me, why are we doing it this way? Reconfiguring, re-explaining the rationale for why we're doing this really helps sometimes to question some of the decisions we've made.

Craig Huhn: I know I remember a few years ago, I hadn't done what we call Algebra C/D now for many years, and people had grown discontented with the direction....We were able to leverage the fact that I hadn't taught it and teach it the first time and say let's just reinvent it, and we start fresh with everybody, what do we want to hang on to. I remember bringing in teachers that had taught it, that weren't teaching it that year like Marty. We said in the past, what are the three important things about parametrics that you think we need to get out of this, and we kind of rebuilt the course—reimagined the course from those sort of purposeful changeovers in who was teaching what and when.

This exchange illustrates the Holt faculty's approach to supporting teachers, both new and experienced. It begins with providing new teachers with not only a reasonable schedule of classes, but also an experienced mentor in the course to support their learning and implementation. It continues with the immediate integration into the Wednesday morning collaborative professional community, with the expectation that they engage with the research and that the decision making is guided by it. And finally, it involves having regular routines in place that guard against complacency and inertia. These ideas include regularly rotating faculty within courses to provide for reconsideration and revitalization of the content, and distributing leadership such that teachers cannot sit on the sidelines of the community, opt not to participate, and decide to teach in ways that they believe are better than the agreed-upon standards of the department.

On Failure

How does Holt make sense of failure and loss? I [Mike] had asked this question specifically because I felt that it was important to understand that while Holt is a success story, that does not mean that everything the faculty touches turns to gold. We noted in Chapter 3, that the path of change at Holt was messy and nonlinear. Failure and loss are a part of any successful organization, in education or outside of it. Excellent organizations have templates for dealing with failure. Apple, for example, has no qualms with canceling a project even after significant investment if it does not meet their design and quality standards. Google chooses to label new and evolving projects as "beta" or "labs," communicating to the end user that quirky behavior and bugs are expected.

While the examples of how technology companies process failure are helpful in understanding a range of possible responses, they also do not translate well to the world of education. In education, we do not invest heavily in a particular class or teacher and then cancel the class or release the teacher midstream if it does not seem to be going well. We can occasionally label an iteration of a class as a pilot, but telling students and parents that they can expect quirky behavior and bugs is a difficult sell when their children's education is on the line. Even changing a simple routine in the middle of a year, such as how homework is assigned and valued, can be challenging for teachers. The discussion that follows regarding how Holt handles failure can be instructive for schools and districts working to make meaningful change.

Dave Hilderandt: I think failure's one of our driving forces. I mean that's what pushes us to make changes. I'm trying to think about what point we realized we needed to do a better job of getting students to write ... there was a district building push [a few years ago], every student had to write, we had a writing prompt, but then when it came to the common exams and the constructed response portion, the first 2 or 3 years we had kids who just wouldn't do anything.

Sean Carmody: Large number of kids who wouldn't do anything.

Dave Hilderandt: And that's totally flipped around. We have very few kids who leave the those [blank. The] number of kids who actually try things has increased, and the quality of response I think has gone up because we knew that they were failing at something so we used that as an impetus ... looking at where kids struggle, where are the kids not communicating understanding. And I think part of it too is that disconnect for me [between] what I observe my kids do on a daily basis versus what I see them do on the final exam, and for some groups of kids there's been a letdown where on the exam they're just not showcasing the skills that I know that they're capable of. And for me I think that qualifies as a failure. So then what do we do, how do we approach that, my changing how the classroom looks, how I communicate to the classroom, getting some new information ... some ideas from my colleagues and other things as a department what do we do in terms of assessing we make students understand that this assessment's important to you. This is something that's valuable for all parts

Sean Carmody: I think there's a couple things there, Dave, that you said that really highlighted some ideas for me, like one of them was that maybe there are certain things that we care about more than others. **That when we define failure, we talk about things that feel important to us** [emphasis added]. Kids answering these constructed response questions, kids thinking they can do math and having enough ... self-confidence to go in and attempt the problem, those things feel like things that we're about in our classrooms, and so that type of failure feels a lot different to me than other types of failure. [We] attack those in a different way, like how do I get better, what is it that's causing this, how do we make this thing help. Like when we talk about math access that's another thing that we... tried to attack like, how do we get more kids in honors classes that are underrepresented, how do we get kids that are predicted right now 100% to fail coming out of middle school, what can we do to get those kids to be successful. Those things to me feel big picture failures that we very much [value and ask,] why's this happening ... is there evidence that this is happening in my room, how do we figure this out, how do we solve this problem. And then sort of dump it back into the group of, well how are you guys doing this? Is everybody seeing this, is this a me problem or an us problem and where do we go from here, and I feel like a lot of my stuff, in terms of those bigger initiatives, is us looking for answers to those questions.

So when I think of one of the things I felt as a failure early on is I had kids that I'm like ok you're successful this year, and I send them off to the next class and they just, mmm, stink. So what did we really do for this kid. And maybe it's a kid that's been unsuccessful his whole life, but he had one year and now he attributes that to you, or attributes that to a fluke, or doesn't and doesn't really take anything with him to keep going. So how do we get those kids to be successful as we continue on. That feels like one of our questions like, why do we have a RAMS class (*mathematics learning support*)? Well we have a RAMS class because we have this population of kids that we're like look, without something else we're just setting these kids up to not experience those big things that we want our program to be about. We want them to embrace reason, we want them to have a self-confidence in math that's different than what they had when they come in, how do we have kids do that? And some of the things in terms of like the loss question, I feel like some of the things that are loss are some of the things we've put in place that I feel like have helped with those things, and are now gone. So we had for a while a RAMS class and it went away. We had the support class and it went away. We had team teaching and it went away, we had these longer periods of teaching and it went away. So there are reasons all of those things happened but for me, the more important reason of how we decide those things is well, what is it doing to these kids to actually allow them to have some access to some things to some choice to be able to graduate from high school to be able to go to college to be able to see themselves as a person who does math, to be able to see that math is something they do when they leave, those are the things that I want us to decide stuff. So when I feel loss is when that's not the way we're making those decisions. [Sometimes] it's a monetary decision, or it's a um, a decision made on something that I don't necessarily agree with is a big goal. So those are the things I don't like. Look, I really want those things to keep happening! So it feels like to me that those losses are the things I keep trying to reintroduce to people to the problems that they're coming up with? Like you know that didn't happen when ... we had that great thing that worked. Hey, how about this idea it was really helpful when we did that.... So I think that those are some of the decisions that we make that drive those sort of [changes], and I think that's how we look for other things [to pursue]. Like, we didn't do formative assessment to do formative assessment. We did formative assessment because we were looking at literature, this makes the biggest impact with our high-risk kids, that's one of the biggest populations that we have, that are trying to do Trig. That's challenging how do we tackle this. So I feel like that was a search to an answer to that question as a group, and I feel like we came up with some ideas and tried some stuff, people had some success and we're like oh, we could do that more. So I think the way Dave illustrated that idea that it was things that are important to us, when we have failure and loss they're kind of similar, like how do we think about this together in terms of a question that's open, and how do we kind of move as a group. And I feel like that's actually, a little bit different than what I remember from *Embracing Reason*. There were offshoots of what people did that were unique to them, but I feel like one of the big differences right now is we're moving a little bit more collectively. And there are advantages and disadvantages to that—we move way slower because we're trying to move as a group. But, we try and move for everybody like that, that's a different thing.

There are so many valuable and interesting perspectives conveyed in this back-and-forth between Sean and Dave. We see in this dialogue a number of important aspects of Holt's protocol of failure. The first step in debriefing failure is determining whether the failure relates to a core aspect of the department's values. If the failure isn't related to something that the department collectively cares deeply about, such as meaningful student learning, improving access and equity, or strengthening the content presented in courses, the deliberation around the failure is likely to be a private matter for the person who experienced it. In cases in which the failure relates to the core values, we see an important first move – naming the problem and bringing it to the group. There is a humility that is inherent in the process that Sean describes. To be able to bring a failure to the faculty community and face the possibility that the failure could be individual (or indeed that it could be something shared) opens teachers up to vulnerability. Yet there is a sense that even if the failure does not appear to be shared, colleagues are likely to have valuable feedback related to the issue. Opening oneself up to the vulnerability has the potential to lead to a shared and meaningful solution.

Sean's statement also acknowledges that failure can sometimes come from outside factors, in the form of removal of a resource or strategy that is working. Even in a highly functional district such as Holt, administrative decisions or constraints can catalyze loss for the faculty. In these cases, Sean observes that those ideas that worked get mourned, but remembered. If and when there are opportunities to bring that strategy back to the surface, those opportunities are leveraged. This appears to be a more functional version of the canon that we sometimes hear from veteran teachers, that every idea comes back around again, so keep those materials from that workshop because they'll be back in a decade or so. The Holt approach of looking for opportunities to bring ideas back to the center, rather than waiting for someone else to do it for them, is an important perspective shift.

The final part of Sean's statement notes an important difference between Holt now and the Holt of a decade ago, in the time of *Embracing Reason*. In the past, there was more variability in the strategies that teachers employed, leading to pockets of difference and (hopefully) innovation in places. At present, the department approaches innovation as a collective, with all teachers planning and trying new things. Sean clearly notes the trade-offs in this decision—change happens more slowly, but there is an explicit attention to supporting all teachers in implementing whatever the innovation might be. Neither one of these approaches are necessarily better than the other—in fact, one may wonder whether Holt would have evolved to the place where they are without having experienced both strategies for change. It is, however, an important caution to be open and explicit about the affordances and constraints of a change strategy, and ensuring that the dispositions and practices of your faculty best match the change strategy you are attempting to employ.

On Shared Values, and Shared Efforts

Shared values can be easy to profess and challenging to live. This is particularly true in education, where the level of visibility into one another's practice can be very low. Holt's hiring processes and community of practice hew to a set of shared values about what is important in teaching and learning. We have seen in discussions about making change and processing loss that these shared values are cornerstones of the decision-making process for the faculty. How do those shared values get fostered, maintained, furthered, and recalibrated when necessary? Dave Hildebrandt and Craig Huhn provide us with a few different pieces of the puzzle.

Dave Hildebrandt: [T]here's inevitable friction, when things don't line up exactly, there's friction. And I like Sean's analogy about when we do get, we made the analogy years ago that a fleet is only as fast as its slowest ship, when we try and move, we're very concerned that when we make a decision to move in a direction that we're doing it together. That we're not leaving people behind. That we're not leaving students [behind], because if you leave teachers behind, you're leaving students behind. And that's just going to catch up to you when students move from one class to another. The habits of mind, the culture in the classroom, the norms the expectations that we have ..., [we've sometimes complained about how are kids are arriving at high school math having] this predisposition towards the math, or were taught that I'm an honors-caliber kid because ... I have good mental math skills, I'm quick to be able to apply any kind of technique or strategy the teacher tells me, the teacher tells me, "here, do this" and I can do it very quickly, I have a great math facility, therefore I'm an honors student. We've been trying to push back against that mentality ... but we know that there's this push we want students to have a certain disposition, and we can't create that from day one when they've had other experiences unless we bring those people, the teachers of those students, and we bring them along with us. Now we're fortunate in that we have a lot of teachers who have gone from our building to [earlier] grades. But I think that's a battle that we've been fighting for a while, it's much easier to get kids to engage with the way we want them to engage with math if they've had practice in the past.

The experiences that students have had when they reach high school have shaped their views of what mathematics is and how one learns and teaches it. Changing that view at such a late stage in students' K–12 schooling can be difficult. This sometimes causes high school teachers to despair, to think there is only so far they can move the needle in changing students' dispositions towards mathematics, or to set aside that effort entirely in favor of "covering the content." Dave's narrative here describes some of the natural impulses in dealing with students who arrive with less-than-desirable dispositions—what is going on in those earlier grades? Rather than engaging in finger pointing, we see the Holt faculty taking multiple approaches to addressing that issue. First, there is a deliberate decision to push back against the mentality that math is about speed and procedure with students, and not just accept it and move on with the mathematics. Second, there's the acknowledgement that a systemic problem must have a systemic solution, and bringing the teachers of those students into the broader discussion is important. Bridging gulfs between buildings can be challenging, and in Holt's case they have had the benefit of teachers moving back and forth between the two high school campuses, the junior high, and the middle school. This migration, by intention or by necessity, can aid in supporting the development of shared values in ways that positively impact the student experience.

In addition to the work of setting norms and targeting dispositions, the mathematical materials in use at Holt High School (in lieu of a textbook-based curriculum) play a critical role in shaping the teaching and learning experience. For years, these materials have lived and evolved on a shared network drive to which all teachers have access. But as we know from studies of curricular implementation, simply having access to high-quality tasks does not mean that those tasks will be implemented in ways that lead to meaningful student learning. Dave and Craig share in this excerpt the ways in which the faculty make sense of the use of those resources, with specific attention to how new and temporary teachers are brought up to speed.

Dave Hildebrandt: Having the shared drive I think has been, and again I don't know what other districts do, how much they share their material ... then again a lot of other districts are working from a preformatted text.... But the fact that we have a shared drive that we [use].... Where we've had the new teachers who were coming in, we've got people who are coming in totally no exposure to our curriculum and they're being asked here teach this whole semester. [We] have documents and materials, we have things that can at least point out where here are some things that we've done, but that's just so superficial. Sean and I would have conversations with [other new teachers], just going over the sequencing and the strategy and the rationale, and like I would work through problems from a student perspective, these are the kinds of things we would've seen, these are the things we'd expect students to have questions about, but that's really the biggest thing, especially when you'll have that experience next year [turning to new hire Spencer], because you have all Geometry?

Spencer Sullivan: Yes

Dave Hildebrandt: Yeah, so part of it's just going through, what is it that we have available to us, what can be used, what's the point, why do we sequence it this way, what are some alternatives.... [S]ometimes I do this sometimes I do this, and we can sit down and have those conversations but again just having that time, and I think Marty and Sean pointed out we've been very cognizant of trying to pair up a new teacher, in a common planning pair with someone who has experience with those courses so we can have that conversation.

(later in the conversation...)

Craig Huhn: I'm thinking about a couple of other shifts, like you talked about the way that we think about our curriculum and the work and [Dave mentioned] trying to organize the shared drive. [T]here are some alterations that are relatively [simple], like font ... and sometimes you personalize it, like *my* dad did this instead of his uncle Phil's oil well.... For the most part though I can identify a couple of big shifts. For a long time if you look back at [older] handouts, they are, do six or eight of these things and what do you notice, so like pattern [recognition], and can you go back later and improve upon, or maybe even prove that pattern? And [now, in part through] Marty's influence, [we shifted to] let's look at the logical necessity for this thing to happen, not notice that it happened and try to retroactively [prove it]. That was a big shift of our assignments. I think there was also a minor shift a few years ago where we tried to descaffold a lot of the things. [Tasks now are] not make the table, now it's just here's a question, there's the thing, so there's been some evolution about how we think about the assignments *en masse* that's not necessarily just idiosyncratic to the particular teacher's font. [That's] one of the things that Marty and I talked about in Boston last year [at a professional workshop] ... when we shifted from sort of the, pattern recognition to the logical necessity approach to the assignments. [So] the way we shifted our view of the assignments that we use, instead of what do you notice here, let's look at seven examples and what do you see, we shifted to the logical necessity approach, and then I mentioned we sort of descaffolded some things a few years back, we purposely put the prompt and then the question and then left it at that.

A collection of high-quality resources is a necessary, but not sufficient, condition for rich, meaningful mathematics teaching and learning. As is evident in Dave and Craig's comments, the resources on the shared drive serve as an important starting point. They have a history, both written and tacit, of how they evolved and how they are implemented. There are unwritten but shared rules about what is fair game for modification and what should remain the same for a consistent student experience. One might wonder after reading these two comments, is there a reason why that history and those rules are not written down? Certainly some aspects could be documented. (In an earlier comment, Dave Hildebrandt shared a story about a failed effort to trace the evolution of the tasks on the shared drive. Despite significant external time and resources spent, the effort proved too challenging to complete.)

I [Mike] would wonder, however, if the oral history component to the resources is in fact a critical catalyst for high-quality instruction at Holt. At the end of his talk turn, Dave mentions the idea that simply giving people access to the resources is not enough—and through tone and gesture, even implied that it might be irresponsible. By leaving the nuance of implementation in the hands and minds of the veteran teachers, it forces meaningful conversations between veterans and novices about teaching using these resources. The documentation of previous experiences and guidance to a teacher could theoretically be done given enough time and energy. But this would increase the risk that a new teacher to the content might interpret that advice in ways that were orthogonal to the intent. In creating an interactive component to the work of learning the curriculum, there are opportunities to understand a teacher's interpretation and gently guide the thinking through discourse.

On Advocacy and Buy-In

The discussion of shared values and shared efforts demonstrates the ways in which Holt supports teachers in embodying a common mission and vision for mathematics teaching and learning. Given the infrastructure that the faculty has put in place with the MSU partnership and what they value in the hiring process, they have created an environment that attracts, retains, and supports teachers who are likely to further that mission. However as all teachers know, even the most like-minded reform efforts among a faculty can be sunk by administrative or parental pushback. Marty Schnepp, one of the longest-tenured faculty members at Holt, provides some insights into how the faculty has approached advocating for their teaching and securing buy-in from the community.

Marty Schnepp: One of the things that I think has kept the pitchforks at bay as much as anything is ... we have people that are passionate about what we're doing and can articulate with empathy and just desire for students to learn more. So when a concern from a parent comes up, 99% of the time that teacher can allay that concern with a conversation because they are so in tune with what we're trying to do, and they can give examples of what their student is being asked to do, and why it makes sense to do it that way instead. I think about trying to have a conversation with one of my son's middle school teachers [in another district], and it was very much like what it felt like years and years ago when I first came here where, you know it's usually this textbook and the teacher can kind of get on board with the parent and complain about the textbook. [In this conversation,] she slapped this binder down almost in hysterics and [said], "Oh my god, look they call this a curriculum there's all this other stuff, like this worksheet and that worksheet..." And then she tried to draw me into this commiseration about the resources they have available, and what ends us happening is you kind of get on board together with that, you're both frustrated by it.

But here I think people are really articulate and passionate about the assignments we have, so when a parent comes with a concern, they can explain what they're trying to get after with that assignment, and it doesn't always work, it tends not to work with the white upper middle class I-want-a-really-good-grade kind of parents, and several other math teachers from other districts. They're the most difficult to deal with because they feel like they've got this expertise and you know they know how to explain stuff really good and get kids to carry out those skills. Administrators can be very difficult in other places too because, ne of the things that became really clear to me when Craig and I were working with that group on that PME [Psychology in Mathematics Education] presentation is that, administrators are having one set of conversations about what good instruction and school-running looks like particular [to] math: textbook publishers and people who want to sell software think that math is just this absolutely beautiful place where it's so simplistic, it's these skills, you just assess them with computer software, you can buy a databank of assessment problems, and that's what they're being told is good practice. And they can analyze that data and do all this, whereas the PME crowd and everybody in thoughtful math education is talking about getting kids to write and explain and justify and do all this other stuff, and those conversations are completely different. And so someone who has dabbled in that world [where it's] all just teach them the skill, drill the skill, I'll do it you'll do it we'll do it, we'll track them with the data software, and everything will be all hunky dory, those people tend to be the most difficult to work with. But when you talk to a typical parent who wants their kid to be in a class that doesn't suck and actually gives them something, in life I think our group is articulate enough to allay those concerns because we're clear about what we're trying to chase down and how that's better for their kid than what they had.

Marty's story about a conference with one of his son's teachers provides a powerful and instructive contrast to how Holt handles parental advocacy. We also see the ways in which the themes we have developed about hiring, shared vision, and understanding of curricular resources all influence one of the most challenging circles of encumbrance, parents and the community. When teachers are living a set of shared values in their classrooms, have the opportunities to make sense of the tasks they are using and how they promote learning, and understand the overarching mathematical trajectory, they are well equipped to have conversations with parents and community members about what they are doing and why. An added benefit of a staff with a coherent vision is that parents will be hearing similar messages across classrooms and teachers as their students move through school. This lowers the chances that a teacher might get singled out by a parent or group of parents for their particular approach. In contrast, we see in the conversation that Marty summarizes a teacher who is clearly frustrated with their resources, feeling isolated, and in trying to ally Marty with these frustrations may be undermining the work of her school and district. This is not to say that her complaints are not legitimate – they may very well be. But what happens when the next semester, Marty's son ends up with a teacher who is enthusiastic about these resources? What messages does this send to Marty as a parent, and to his son as a mathematics learner, about the work of doing mathematics in that district?

To summarize, the first and perhaps most critical aspect of advocacy is having a strong rationale for what you are doing and the ability to articulate that rationale in a reasonable way to parents and community members. The second aspect is that a strong faculty that engages in regular and meaning-ful discussions about teaching and learning helps develop, refine, and grow that ability to describe what you are doing and why you are doing it – and to advance that work individually and collectively.

We realize that as you read these words from the Holt faculty, they may engender feelings of frustration, and perhaps even despair, if you are in a building and district that does not operate in these ways. We did not want to end our sharing of teacher voices without asking our veteran teachers what advice they might give to someone getting started with this work. Each of the teachers in this discussion came into some aspect of the work, depending on when and how they arrived at the district. That said, each of them have played important roles in advancing that work through their teaching, leadership, and research.

Advice for Revolutionaries

You picked up this book, we expect, because you want to make positive change in the teaching and learning of mathematics in your school. Maybe you're on an island and starting this work just in your classroom. Maybe you have some willing accomplices already in your building and are looking for ways to gain momentum and advance your collective work. It is our hope that the closing perspectives in this chapter, advice for teachers looking to start similar work, meets you at whatever place you are at and pushes you to consider new and exciting ideas to advance your practice.

Marty Schnepp: I think about this group that Craig and I have been working with in Boston, [the] Better Math Teaching Network. It's spread out across New England, and they didn't want to be alone in their school, to a large extent. So people are finding ways to network across fairly large dis-tances, because it's difficult when you have a static entrenched environment to make movement on things. So the only thing I can suggest is what I know, go to the Park City Math Institute or the Better Math Teaching Network, or somehow find a group outside that you can get going with, and then finding ways to engage more thoughtful col-leagues in what you see kids doing or not.

You know I go back to an overtold story, back at my first school in San Diego where, MSU was the academic learning program back then, and there was talk about what's going on in your room and how to get better at it. And, so I started asking other people how they approached a particular topic at lunch and at the end of the day the principal's standing outside my door and when the kids let out she pulled the door behind them and said, "I hear you're struggling." And, that's the way the culture of a lot of places still 30 years later. So I really empathize with somebody who is going it alone. One of the things I feel like might be possible in places like Boston ..., places where schools are being created and dismantled year in and year out, trying to find a way—like in Chicago I've got some friends who all finagled ways to get to the same school. So they have built a group of people together that have a common vision of trying to get better at stuff and there's something more that we can ask kids to do rather than just grunt their way through [the math]. And so I think if you are alone, just do what you can to not be alone because you can't go it alone.

Sean Carmody: It feels like there are some other things too ... one of the things that I would encourage people is, you need to do stuff in your room to get better. Like, you need to try stuff, you need to be working on some things. Because it feels like finding those extra people is important, but if you start working on that you could find the people in your building, whether they're in your department or not.

Dave Hildebrandt: I think one of the things that we have always done is that we've always tried stuff, we don't know if it's going to work or not but **we never, ever, have been satisfied with just not being working on something, not trying something new** [emphasis added]. If it's a reading, if it's a group study thing, if it's implementing something in the classroom, collaborating, we're always wanting to try stuff because you don't know what's going to work until you actually implement some changes. We're never satisfied with what we have. I can't imagine many of the teachers here would be satisfied with

> the status quo, I think that's been our driving
> impetus, we always know there's something better
> we could be doing.

Change starts, but does not end, with what you do in your classroom. Create a culture for yourself of not being satisfied with the status quo and finding something new to work on. Keep a list of the aspects of your practice you want to further develop. When you have worked one aspect of your practice for a while, move on to another and work there. Hang on to the notion that change is dynamic and iterative, and that one can always return to a previously-worked aspect of practice at a later time and develop it even more.

Finding colleagues can be more challenging, particularly if you do not have an immediately clear set of allies in your school. But do not be satisfied with this state of affairs, and look for opportunities to take advantage of that. Conventional networking opportunities like local and state math professional organizations are excellent places to begin. More intensive events like the Park City Mathematics Institute and PROMYS for Teachers are also good opportunities if you or your district can help to foot the bill (note: at the time of this writing, both of these programs either fully or partially subsidize costs for participating teachers). The ubiquitous connectedness of our 21st-century world has also created highly active online communities like the Math Twitter Blogosphere (#MTBoS) that can provide both substantive and palliative support nearly in real time for mathematics teachers at all levels. Interact with people, build trust and shared values with them, and then recruit them to your building when the opportunity arises (or allow yourself to be recruited to a place where you are a better fit).

YOUR TURN: RADICAL CONVERSATIONS

The conversations that we curated among the Holt faculty helped to recreate and revisit the curricular and pedagogical history of the high school. The whole was greater than the sum of its parts—no one teacher we spoke with had a complete and fully articulated recounting of their years in the district. As stories wove together and memories were filled in, we were able to form a rich and coherent narrative about the key moments of change at Holt High School. You have just finished reading that narrative here.

For your Radical Conversation of this chapter, we would like you to hold a similar discussion with your faculty. This will build on the conversation you had in Chapter 2 where you laid out the timeline of events related to curriculum and policy. It builds on and broadens the conversation you had in Chapter 3 where you engaged your colleagues in a discussion about how superficial changes relate to deeper pedagogical issues.

You can have this conversation in a relaxed environment—on a weekend over brunch, on someone's patio over beverages, or as a part of another social gathering will help people feel relaxed and willing to share. The goal of this conversation will simply be to construct a shared history that is richer than any one person's contribution.

Start with the following question to structure the discussion and connect to the previous ones:

- What impact has the changes that we have made have on our teaching, and on student outcomes? (The Radical Conversation in Chapter 5 will ask you to follow up on the issue of student outcomes.)

Take notes during this conversation so that you remember the key points. If your colleagues are amenable, you might even consider audio recording the conversation to be better able to represent the key points. Following this discussion, dig into some or all of the following prompts, inspired by the conversations with Holt faculty:

- What structures and routines do we have to make changes in our programs?
- How do we handle changes that come in from the outside?
- In what ways do we build shared values?
- How do we handle failure and loss?
- How do we message to parents and the community what we are trying to do with our mathematics program?

Leave at least an hour total for this discussion (two hours is probably more reasonable). Share the documentation of this conversation with your colleagues when it is complete. Encourage them to add to or annotate that document if there are ideas that they didn't feel were fully represented.

INVESTIGATION AND REFLECTION ACTIVITY 4

Your Investigation and Reflection Activity involves finding accomplices.

Choose one of the areas below in which you would like to see your building and/or district make change. For that area, find both some written resources that will help you (like tasks, course plans, or research) and some human resources that will push your thinking. Choose one of the following areas to explore:

- How to change the content that we teach in our course program and to whom we teach it (to promote more equitable access).
- How to eliminate tracking/ability grouping and still meet all students' needs.
- How to create a long-term agenda for change (to both content and pedagogy) that moves towards a more student-centered mathematics program.
- How to better measure student success over time (beyond grades and standardized assessments).
- How to create advocacy tools to help administration and parents understand a more student-centered approach.

If you are not sure where to start for written resources, position statements and research briefs from organizations like the National Council of Teachers of Mathematics, the National Council of Supervisors of Mathematics, the Association of Mathematics Teacher Educators, or TODOS: Mathematics for All. (Position statements and research briefs are available on their websites.)

If you are not sure where to start for human resources, contact your grade-band representative with your local NCTM affiliate (e.g., Michigan Council of Teachers of Mathematics and Wisconsin Mathematics Council for Craig and Mike, respectively). The Twitter #mtbos and Facebook groups focused on mathematics teaching may help. Your regional education agency of state department of education is likely to have a mathematics specialist eager to connect and talk with you. And finally, a local university's mathematics educator would likely be willing to provide you with conversation and guidance.

After you have identified these resources, make a short- and long-term plan to move forward on the point that you have chosen. This can build on the idea that you considered in Chapter 2. Make sure to include in that plan how you will be making use of those resources for help and support. You do not have to do this work alone.

CHAPTER 5

MAKING A DIFFERENCE

Student Outcomes at Holt High School

The ultimate goal for any secondary mathematics program ought to be to create a generation of future citizens who recognize mathematics as worth knowing and a smart way to make decisions; that can use math they know to solve problems in real life; and have the perseverance, curiosity, and confidence to create mathematics for problems that do not yet have known strategies. This puts us as teachers in a position of living the values that research has so clearly communicated for years—that experience with nonroutine problem solving is the key to enduring, meaningful student learning in mathematics (e.g., NCTM, 2014; Stein, Smith, Henningsen, & Silver, 2009). The problem, though, is that measuring the effectiveness of your program on this lofty ideal has practical limitations. Usually, the best that schools are willing and able to do is check that students are able to reproduce the mathematics they have been taught. And ironically, the most trusted or used metric to judge and compare schools is a national test that is based on how many of your students you get to perform correctly on a timed test of traditional content-related items, as noted in prior chapters. One needs to look no further than the rise and dramatic fall of the Smarter Balanced Assessment Program (Rasmussen, 2015) or Mike's story in Chapter 2 about the Maryland State Performance Assessment Program to understand the challenges of establishing a conceptually-oriented assessment that measures the ability to think and problem solve on a seminational scale.

A Quiet Revolution:
One District's Story of Radical Curricular Change in High School Mathematics, pp. 93–113
Copyright © 2018 by Information Age Publishing

At Holt, we still obviously test for understanding of the mathematics content. However, we try to make sure our assessments require that students understand the ideas, and not just remember the punch line. We ask them to perform on questions that they have not necessarily been trained to answer, in the hopes that students with superficial procedures-and-memorized-facts knowledge are distinguished from those who can justify, explain, apply in nonroutine contexts, and find the mathematics in problems where they do not immediately know what to do. And rather than make this the line between an A and a B, we have the stance that this ought to be the line between passing and maybe failing. A student who can only perform procedural tasks without links to big ideas and connections throughout will likely not be able to earn the credential of passing. And, this does mean that some students have to retake the course a second time to earn credit.

You might have noticed in the previous paragraph, we used the phrase "maybe failing." This is a reasoned turn of phrase, as committing to these standards of mathematical achievement goes hand-in-hand at Holt with committing to the notion that every student can achieve these ideas, but perhaps just not in the time frame we might hope. Maintaining this growth mindset perspective on student learning (Dweck, 2006) requires us to reframe what it means to teach and learn and how that relates to course completion and the ultimate granting of Carnegie units towards a diploma. In an effort to destigmatize needing more time to become proficient on our high expectations, our department has created a "not yet" credit option for students with consistent effort who were still unable to earn credit. This rewards them for working hard by awarding them an elective math credit (as if they audited the course) instead of an "E" on their transcript and GPA. Then, they re-enroll in the course and most times, become class leaders the second time through. What we have learned, though, is that this situation is pretty rare—it is uncommon for a student to show consistent effort and still not quite demonstrate enough understanding needing more time. Nevertheless, we find it an important component of our shared values to find a way to support those uncommon students for which more time leads to the understanding that we collectively seek.

The challenge for us at Holt is always how to claim that what we do makes a difference. How do we know that students are coming out of our program with anything different in terms of their mathematical knowledge, skills, and understandings as compared to somewhere else? I have always desired a scenario where a handful of high-achieving students and a handful of middle-achieving students from all area schools would be given an assessment that can determine who "really get" algebra—by which we would mean understanding the conceptual underpinnings and being able to tackle nonroutine problem solving in the ways advocated for by educational research and NCTM (e.g., Chazan, 2000; NCTM, 2000, 2009). The closest we have come to that was a project I worked on several years ago now where we studied how students with special needs from area districts made sense of mathematics (Huhn, Huhn, & Lamb, 2006). What we found then was that, among a small number of students with special needs, there was a tangible difference between the Holt students and those from outside of Holt. But this one-time observational study is far from sufficient to make any broader claims even if we wanted to do so. But ultimately, getting students to do really well in mathematics is not a zero-sum game. In other words, we really do not care if we are outperforming other districts on the truly important indicators of mathematical proficiency; in fact, what we care about is more kids understanding more important mathematics regardless of district. In fact, that is the stated goal of this volume—that we can share what we have done in our district in the hopes that some will be useful across any and all districts looking to continually try to improve.

So what have we looked at to evaluate whether the changes we have made have been working? It is worth noting that any time we try something in our department, we hold ourselves to a high enough standard to be able to have evidence. For example, we pulled the data for students who were re-taking Algebra to confirm our observations that they seem to be successful the second time through (their success rate was indistinguishable from the general population). When teachers share fear that having students retake a class will throw them into a spiral of despair, we can share what we found that having students retake the course again got us a much better success rate for that group of students than we had with most other interventions we have tried. But for the purposes of this book, we are speaking more generally—at a macrolevel, what student outcomes can indicate that students are learning to appreciate and use mathematics at a deeper level than the way most of us memorized a series of procedures?

The crux of the argument is that, due in some part to asynchronous and serendipitous pronouncements from the State of Michigan, the willingness of the Holt High School (HHS) math department to engage in the work, and the support of district administration and close ties to Michigan State University's Teacher Education program, Holt has evolved to a place where a decently large district has essentially all of its students graduate with mathematics up to and including Precalculus. As described in Chapter 3, the gradual evolution of the high school curriculum through content shifts within courses, school configurations, and graduation requirements has led our faculty to a place where we are able to successfully support our students in completing rigorous mathematics beyond the traditional minimum high school endpoint of Algebra 2. Thus, a high school has quietly restructured its courses and adhered to the state's graduation requirements so that in order to earn a diploma, all students must demonstrate true understanding of mathematics up through a Precalculus experience that includes substantive treatment of functions, trigonometry, and probability and statistics. This also means that if no course needs to be retaken, *every* student at HHS has access and an opportunity to take AP Calculus and/or AP Statistics. This is in contrast to the standard practice of having many lanes or tracks for students of different perceived levels (see the anonymous course sequences from real districts in Figure 5.1), or having the highest course required to graduate be a traditional Algebra 2 that would still require a course like Precalculus afterwards. At Holt, all students take Geometry, then 2 years of Algebra, and then if they did not repeat a course, a math elective their senior year. It is clearly evident in comparing the Holt options to the three peer districts in Figure 5.1 that not all students are afforded the same access to Precalculus content or beyond. For example, in Illiwauk, only students who complete Algebra I in Grade 8 have access to Precalculus without doubling up courses. In Meadow, students who have struggled with mathematics prior to high school have no access to the content at all. Elm is similar—unless you have taken some form of Algebra in eighth grade, your only option for accessing Precalculus is to take five mathematics courses in four years (and be successful in all of them). Students who enter eighth grade in the lowest of the tracks would have no reasonable path to content beyond traditional Algebra II without doubling up. What is more is that all three of these example course sequences segregate students in three or more tracks, and mobility between those tracks is possible in theory but limited in practice. Elm explicitly represents this with their dotted lines. Meadow provides no upward access to their honors track until Calculus. These structures implicitly embody a fixed mindset approach to mathematics learning, and privilege performance at the middle school (or earlier) as the primary predictor of the mathematics students will study in high school.

To further the point, being able to earn credit in the required courses to graduate at Holt requires an in-depth understanding as well. This is in contrast to tests we have seen from colleagues in other districts that can allow students with rudimentary knowledge or memorized procedures to "get by" with low-level questions on tests, often with formulas and definitions given. Often times, districts go so far as to ensure that a passing performance on assessments with a C is possible if students can answer only the procedural questions correctly, positioning problem-solving and application as superlative and not strictly necessary for adequate performance. As mentioned above, we try to have assessments that check for understanding and require explanation, as well as requiring students to demonstrate they are able to perform mathematics with the complexity required of the course expectations. For example, in our first-year Algebra course, the Polynomials Unit test asks students to explain the least number of x-intercepts a degree 10 polynomial could potentially have and why

Illiwauk District Mathematics Course Offerings

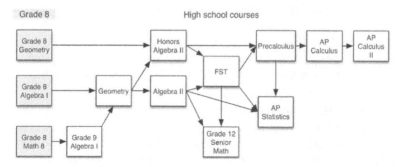

Meadow District Mathematics Course Offerings

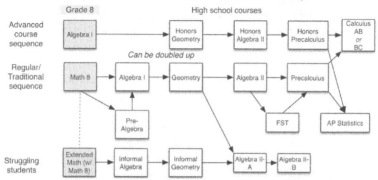

Elm District Mathematics Course Offerings

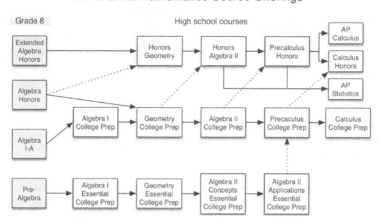

Figure 5.1.   Sample course sequences from Holt Peer Districts.

to help us determine who can really understand the end behavior for even-degree polynomial functions. While degree 10 polynomials are very rarely in play even in careers that make use of mathematics, considering this question with respect to such a polynomial requires students to fundamentally understand the structures of polynomial functions, what makes them predictable by family, and translate that to key mathematical features across function families like x-intercepts and end behaviors. Later in the Rational Functions Unit, besides asking students to show all of their work to find solutions to $\frac{7}{5x} = \frac{x-8}{x^2+6x-4}$, we also require they find the discontinuity for a given rational and explain why this function explodes to infinity as you choose values closer and closer to it. In the Periodic Functions Unit of the second-year Algebra course, we demand students can write an accurate rule for a given sinusoidal graph, and also then be able to find the infinite number of solutions for a given y-value. For the four semester final exams in the Algebra sequence, part of the final includes two constructed response questions that gives a scenario, and requires the students to completely determine what they need to do to come up with a reasonable answer (and then, tie it back to the situation), linking the understandings of functions to the work of mathematical modeling as conceptualized in current policy recommendations (Consortium for Mathematics and its Applications & Society for Industrial and Applied Mathematics, 2016). These are graded together by the department on a rubric that includes both process/explanation and accuracy components.

Besides just expecting students show they understand the big ideas and can do the important procedures, our department has worked really hard to adjust the grading procedures to ensure that students earn their credentials through their knowledge and not through "easy" grading practices. Even prior to the very popular work about Formative Assessment and grading practices that is currently in vogue in the professional literature, such as the importance of formative assessment (e.g., Black & Wiliam, 2010) and the problems with grading (O'Connor & Wormeli, 2011), we were thinking about making sure grades truly represent the amount of learning. See, for example, "How Many Points is This Worth?" in the November 2005 *Educational Leadership* (Huhn, 2005) where I made the argument that the way teachers often think and talk about grading has been often self-defeating. Soon after this, a colleague—Sean Carmody—suggested that a group of us in the math department work over the summer and really study this formative assessment stuff (recall the narrative about the evolution of this initiative in Chapter 4). A subset of the department then read books and articles together and talked about how to implement changes in our classroom; a conversation that has spread now to not just the full department but across the building as well. We even have several cross-curricular groups in our building working on formative assessment practices like organizing their courses around clear learning targets and giving actionable feedback to students, and teachers are experimenting with standards-based grading at the high school level. Through all of this, our department has come to the point where grades are almost solely or even completely based on the understanding demonstrated on individual assessments of the mathematics at hand.

Ultimately, the goal of our program is to be able to say that for a student to earn a diploma from HHS, they need to be able to really prove that they have gained a considerable amount of knowledge about Precalculus level mathematics, and we do not let them off the hook. And even though by state law, students with disabilities are allowed to option out of the second year of Algebra, we have very few students take that route. We achieve this partly through a decades-long commitment to equity though the district's immersion philosophy for students with disabilities or other learning challenges. Students with disabilities have been members of their classes since elementary school, and students rarely bat an eye when special considerations arise. Through a deal reached between administration and the teacher's union long ago, students with IEPs are counted as two when filling classes, allowing a little extra time and space for the teacher to account for their needs. And, our building works really hard to combat the notion that only some kids can learn mathematics. For some of those outside the department, it has taken reminders that learning mathematics is the same as learning to read, drawing attention to and contrast with narratives that it has become socially and professionally acceptable to think of matheamtics as somehow genetically predisposed as a path for the few. This ethos, and the consistent press to help all students and teachers understand it, preceded the current and very popular research related to growth mindset, and the hope is that the additional prominence of that work will bolster the equity that we have been fighting for at HHS for decades.

So we created a high level of expectations for all kids and innovated some supports for students who may need more time. But, I believe, one true secret to making this all work has been the ongoing close relationship with MSU, who has been the top secondary teacher education college in the nation for decades. We have utilized them and the connections made through them for professional development, projects, and hiring. The reason that there has been moderate success in what we have gotten kids to do is because our teachers, by and large, can offer high quality classroom experiences. We do not have a textbook, and so the teachers we employ need to be thoughtful, flexible, inventive, curious, and thoroughly grounded in pedagogical methods (see Chapters 3 and 4 in Chazan, Callis, & Lehman [2008] for insights into the curriculum and planning processes). And, because we often host interns from MSU, we often can train prospective employees for a year of guided collaboration. This gives us not only a sense of a prospective hire's pedagogical skills and content knowledge, but also provides a clear window into their dispositions towards teaching and learning mathematics (along with an opportunity to move the needle on those dispositions). Again, this places the work at Holt at the forefront of what is now a national movement to explicitly consider the ways in which we support teacher candidates in developing positive dispositions towards mathematics and students (Association of Mathematics Teacher Educators, 2017). The professional development side of the partnership provides an external sounding board for innovation, a conduit into the district for new and exciting educational research and development, and opportunities for teachers to hear an "outside voice" that reinforces the work when such a voice is necessary. Not all schools have the advantage of such a partnership; however, the goal of hiring candidates that are consistent with best practices, having high quality professional development, and a collaborative team of teachers can be replicated through sound hiring practices and strong leaders. As we elaborate at the close of this chapter, we feel that there are steps that districts who do not have such a partnership (or even the possibility of one in close proximity) could take to approximate or replicate these conditions.

What we have been able to create, over many years of incremental change, is a requirement beyond geometry that all students actually learn polynomials, rationals, exponentials and logarithms, sine and cosine, probability and statistics in order to earn a diploma from a large urban-adjacent high school. These topics are learned with meaning in ways that are ruthlessly authentic to both the mathematics and the utility thereof. And, contrary to the larger social belief that math is not for everyone, the district did not implode. In fact, the graduation rate for HHS has been essentially constant, even as demographics have shifted and these changes have been evolving to the place we are now. Let us spend a bit of time digging more deeply into these data. We acknowledge the substantial limitations of large-scale testing as a measure of student learning, and as such, we make use of national, state, and local data to paint a more nuanced picture of student performance.

THE DISTRICT DID NOT IMPLODE: STUDENT LEARNING OUTCOMES OVER TIME AT HHS

A look at large-scale assessment outcomes. One conjecture we always had was that even if we do not believe that the large-scale standardized testing can tell us how much understanding of mathematics our students have, that if we taught more students more mathematical stuff, that it would have to show up on things like the state-mandated ACT or now SAT. And from

Mike's perspective as someone who conducts professional development across a number of districts, my stance has always been that supporting students in deeper mathematics thinking and reasoning would translate into good test-taking skills that would allow them to reason successfully through more procedural items. (See Chapter 7 and Appendix for more discussion of the authenticity of assessments like the ACT.) As it turns out, when you avoid teaching tricks, take more time to deeply delve into big ideas at the expense of "topic coverage," and honor contemplation over speed, it does not show up in obvious ways. However, with some closer inspection, we did notice something interesting. In 2014, while looking at the state accountability data (at the time, the Michigan Merit Exam [MME] was for all high school juniors and made up of the ACT, the ACT WorkKeys, and a special Michigan addendum to account for Michigan standards not assessed on the ACT) we saw a trend that surprised us, and when we compared it to other schools appeared to be uncommon.

What we saw was that for the prior 2 years we had data for (2011–2012 and 2012–2013), there appeared to be a near universal drop in math proficiency as students progressed through the middle grades and into high school for a variety of districts (high or low performing, small or large, higher or lower SES, etc). But, at Holt, we seemed to buck that trend.

Table 5.1.
Comparing Holt and State Averages On State Assessments

		2007–2008, 7th grade class	2011–2012, 11th grade class (same cohort)
Holt data:	Holt MEAP mean scaled score	716.6	
	HHS MME mean scaled score		1103.2
State data:	MEAP mean scaled score (SD)	717.1 (27.4)	
	MME mean scaled score (SD)		1098 (33.6)
Holt's *z*-score *(state mean–Holt mean)*/SD		–0.02	+0.15

The data in Table 5.1 shows in general, Holt students performed rather close to the state mean in assessment data for both assessments. In looking at the change over time, however, this cohort of Holt students went from –.02 to +.15 in mathematics (a gain of .17) as they went through our secondary math program. When we checked, many other schools did not have that same experience. In fact, for the schools in our area (plus one on the west side of the state that we are often compared to):

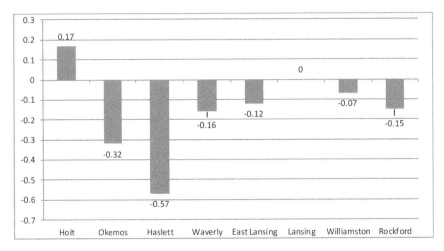

Figure 5.2.   Holt z-scores compared to other districts in the region and state.

When we repeated the analysis a year later for the 2012–2013 data, Holt went from a z-score of .02 in seventh grade (in 2008–2009) to .10 in 11th grade, for an increase of .08. With updated calculations for the new scores in other districts, we again see the same trend as before. Rockford, for example, went from .53 above the seventh grade mean to the same cohort being .30 above the mean as juniors (a change of –.23). These negative trends between middle and high school are consistent with data from the National Assessment of Educational Progress, which has shown an upward trend in eighth grade scores since 1973 while high school scores have remained flat since that time (National Center for Education Statistics, 2009). This helps to strengthen the hypothesis that something promising is taking place at Holt, and not that the surrounding districts are somehow anomalous in their declines in student performance.

So the argument is that, even though we do not believe the state testing data says much about the level of mathematics that students really understand, and certainly does not assess the kind of mathematical habits we work to instill in our students, for skeptics we can at least see if we are getting our set of kids in the middle grades from where they are at to perform better by the end of 11th grade. It is noteworthy that this is a large-scale overview: the set of kids in a graduating class in 7th grade is not the exact same set of kids in the graduating class by 11th grade; and even if they were, this kind of test may not even be comparable (although both the MEAP and MME are similarly procedural in nature). But, for those that worry dismissing standardized tests out-of-hand can diminish the argument, we have been able to show groups of students increasing their performance relative to the rest of the state as they move through our sequence on the preferred measurement. And for district administrators less knowledgeable about best practices in mathematics and fed in their admin classes lines stressing the importance of "data" and "checking the health of the system," arguments like this—using the data that *they* value—can provide the freedom to continue to work on practice in the direction we know is worthwhile.

Zooming in to the local assessment level. One of the ways we try and ensure we offer an experience across the board for students that will allow for an increased likelihood of true mathematical learning is, as previously mentioned and chronicled in *Embracing Reason*, a level of collegiality and collaboration coupled with time carved out for in-house professional development. At that time, the State of Michigan would "forgive" a certain number of embedded professional development hours to count towards student contact time, making the argument for protected time as a mathematics faculty much easier to make. In the intervening years, the number of hours eligible for forgiveness had dropped to zero, so we have fewer late-start Wednesdays than we used to; about three per month rather than four. But one of the other quality control measures we have put into place is a shared ownership for how our students perform on final examinations.

Our final exam is made up of three parts: a Constructed Response (CR) portion, a Short Answer (SA) portion, and a section with Multiple Choice (MC) (a vestige of the principal that associated multiple choice performance with stronger standardized assessments scores noted in Chapter 4). The Short Answer section is our chance to ask them to perform or explain, where we can score their responses on a department-wide rubric. Between the SA and MC sections, we feel we can get a really realistic view of what the students understand and can do across the content for the semester. But, we also want to know that the way we work with our students allows them to see the utility of the mathematics, find the math in a real-life situation, and then tie the mathematical answer back to real life. Therefore, we ask them to do a couple of constructed response questions, where they are given the scenario and a question about that story, and then are asked to show how mathematics can help the person come up with an answer. This is in contrast to the Short Answer questions, where we direct them what mathematics we want to see; for example, "Solve for when $f(x)$ equals 9," "Convert the following into vertex form," or "Explain why you cannot find an inverse rule for $9x^3 - 4x^2 + 3x - 10$." For the Constructed Response, they need to *find the math* (determine the variable they want to use, write rule, determine if they need to evaluate or solve, etc.) and the task design and our rubric allows for multiple approaches. The goal is that by the end of each term, every student is comfortable approaching problems that they do not immediately know what to do and can create a mathematical response that justifies a realistic answer and communicates their thinking well to the reader.

The rubric we use for these problems has two components: a process portion (did the student know what to do?) and an accuracy portion (did they do it correctly?). For process, we are largely looking to see if the students approached the problem in a way that would lead to a useful solution, if they communicated their strategy well, and if they interpreted the answer they got in relation to the context (stategic rounding, appropriate units, disregarding superfluous solutions, etc.). For accuracy, we are interested if they performed the mathematics correctly and found a realistic solution.

What we do then is split the exam into two parts and have the Constructed Response section (two or three big questions) on the last day of regular class, and then the other two sections during the exam period itself. That way, students have just about a full hour to focus on writing a few good, thoughtful responses. But it also serves a pragmatic purpose—having the Constructed Response section completed prior to the exam period itself means that we can, as a department, grade all of the students' CR problems together. We have about 550 students in the first year Algebra course, and so the teachers of that course bring all of the CR booklets together, and we split into teams of Problem 1 and Problem 2. Ideally, the Problem 1 people then grade all 550 responses together with that common rubric, and similarly for the Problem 2 team. This gives us consistent and reliable grading, but more importantly it gives the teachers of that course a wide view of what students are doing. Often we will see patterns or generalizations that we can talk about and inform our approaches. Speaking from experience, there is considerable self-assessment as you compare what your students do to everyone else's students. It constantly generates conversations about what teachers did in their classrooms or how they reviewed content.

This common grading of finals, along with the long-standing collaboration and culture of the department bolstered through professional development time built into our calendar, is one vehicle for not only reflecting on our own practice, but having constant conversations about the department goals and what we want to see students doing. Working together like this, much more than any external accountability, has generated the most introspection and desire for continuous improvement, both individually and collectively. Activities like this represent good collaborative teaching practice and serve to reinforce the standards, ideals, and values that we share as a department of mathematics. More pragmatically, a common final serves as internal quality control and, we believe, does a good job of assessing true understanding of the content so that we can then ensure that our students have a consistent, high-quality experience that leads to improved student outcomes that we value.

A broad view of student achievement. When *Embracing Reason* (Chazan, Callis, & Lehman, 2008) was released a decade ago, we were not claiming that we had all the answers or that things were amazing at HHS (as is the case still with this volume). We were, however, trying to chronicle some of the changes made to try and improve the instruction we were offering. Among the highlights of the data analysis at that time were the following as shown in Figure 5.3.

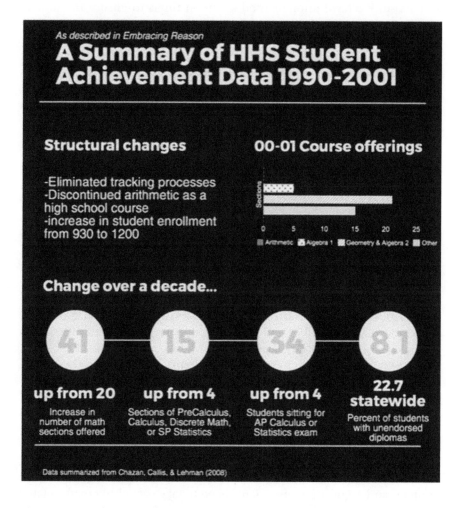

Figure 5.3. Achievement data and shifts in Holt High School mathematics, 1990–2001.

Since that time, we have continued to make changes and evolve; and our hope is that this change has meant even better experiences for students than "the good ol' days." For example, we have reduced the impact of tracking that still lingered and impacted course schedules a decade ago in our building. Now with a revamped middle grades curriculum, we have all ninth grade students in Geometry (or the Honors option) and all sophomores in Algebra (or the Honors option). We have also maintained the ability for all students to elect for themselves the Honors option if they so choose. We have implemented common finals to ensure the understanding at the end of our courses is consistent across the program. The district has begun in the last 2 years to have a K–12 conversation about a district-wide vision of mathematics learning in line with the vision of the National Council of Teachers of Mathematics. Our department has utilized the "not yet" credit idea for those students who worked hard but, in cases of no fault of their own, were unable to become proficient enough to earn credit. We have experimented with different forms of supports for struggling students. We have chosen to continue and expand the functions-based approach to algebra as described in *Embracing Reason*. We have also, as a district, chosen to continue and value the embedded professional development time on Wednesday mornings, even though the amount has been slightly reduced in response the changing state laws. In the changing demographics of the district and political context of public education, it is hard to be able to say definitively what impact would be expected through these deliberate decisions. It is always problematic to measure success in human services like education without a control group; and there is danger in setting a "good enough" threshold of success—if is below 100% you are either saying not all students matter or inviting complacency.

One of the metrics that *Embracing Reason* examined (Chazan, Callis, & Lehman 2008, pp. 109–121) as evidence was the number of students who elected to take math courses beyond the requirements. At the time, the graduation requirement was passing two math classes, and these math classes could be at whatever level the students entered based on the tracking practices at the junior high at the time. So looking at the number of students who continued in their math path when they no longer had to was one way to see if teachers and courses were having an impact on students' views of what mathematics is and if it is worthwhile, or even enjoyable, to know. Now, the State of Michigan has set graduation requirements at Geometry, Algebra I and Algebra II (with some exceptions to the latter) content, and requiring that students need to take something math-related in their senior year. Therefore now, there is much less room for students to elect math beyond state requirements, although the definition of math-related has been loosened significantly. Also, with the higher expectations and requirements for all students, students sometimes need to retake one of the required courses. As a result, comparisons are not an exact science, and attempts at some claims might be viewed as spurious. However, with a 4-year graduation rate over 90% (mischooldata.org, 2017), Holt High School has nearly every one of these students (minus a handful of special education students with a legal exception to the state requirement) taking and passing a rigorous Precalculus-equivalent course studying trig functions, probability, and statistics. In an era in which students can take on online music theory class to fulfill the math in the senior year requirement, and less than half the students go on to a post-secondary education, 267 students (out of about 450) took a course beyond Precalculus at Holt in 2015–2016. These courses include Advanced Algebra Topics, Data (a course that covers about the first half of AP Statistics), AP Stats, and AP Calculus.

Accessible grade records only exist back until 2007–2008, but at that time we had 383 juniors in the course that studied the same topics (at the time, it was called Functions, Statistics, and Trigonometry). Out of a total of 454 juniors that year, 365 of them successfully earned credit (80.4%). In 2015–2016, we had 386 juniors taking and passing the course (now called "Algebra C"), out of 423 juniors (91.3%). So we have more kids learning high level mathematics now, with oversight on rigor and a much larger diversity of students, than we did even during the decade that *Embracing Reason* was published

Additionally, the penultimate outside assessor of quality, the Advanced Placement tests, corroborate this increased level of quality for Holt's students of mathematics. *Embracing Reason* (Chazan, Callis, & Lehman, 2008, pp. 117–118) references the growth in the number of students taking the AP tests in mathematics (as well as the rise in the total sections of students taking the courses) and their success on the exam. According to *Embracing Reason*, in 2000–2001, 29 students took the AP Calculus test from Holt, and 14 took the AP Statistics one. In 2015–2016, 50 students took the AP Calc test and 33 took the AP Stats test. And this larger group of students learning more mathematics is demonstrated by the number of students receiving a top score: in 2000–2001, eight students at Holt earned a five on either the AP Calc or Stats test. In 2015–2016, 18 did. Many people are surprised to hear that Holt, as a working-class community, can get 132 students (in 2015–2016) to take an AP course in mathematics. But as we have argued here, if you can convince students that learning math is doable and worthwhile, and avoid systemic barriers through tracking, students will follow. Figures 5.4 and 5.5 show a comparison of the 2001–2002 data to the 2015–2016 data for both courses.

When we tell people that every student at Holt has the opportunity to take an AP class in mathematics as a senior, we are often met with blank stares or confused looks. But the reason is simple—we have one path for all

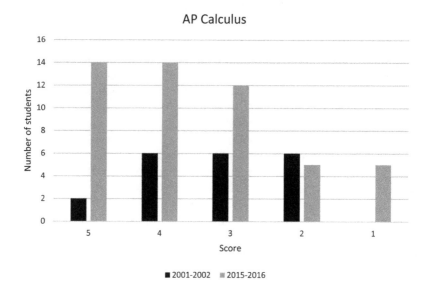

Figure 5.4.   Advanced Placement Calculus Data, academic years 01–02 and 15–16.

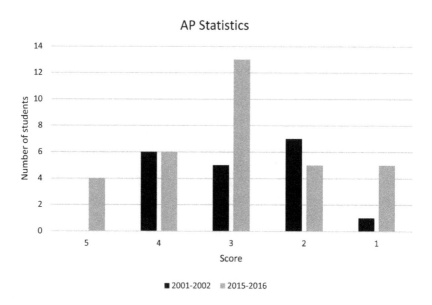

Figure 5.5.   Advanced Placement Statistics Data, academic years 01–02 and 15–16.

students to take (albeit with a choice to take Honors version or not along the way) where when completed successfully, anyone is prepared to take the AP courses as a senior of they so choose. It begins with every student being placed into Geometry as a freshman. This allows for a clean slate regardless of success in junior high (or their previous district as move-ins and school-of-choice students tend to enter at this point). The students and parents decide if they would like the regular course or the Honors version, which has additional curricular topics or a deeper focus on some for students who see themselves as college-bound. Placing Geometry in ninth grade has two important purposes—it maintains an uninterrupted flow of the connected topics between Algebra 1, Algebra 2, and the senior options; and also because that provides us some concentrated time to intervene with students who struggled in the pre-Algebra work in the middle grades.

For the students who failed seventh and eighth grade math, or are flagged for certain indicators based on unit scores for prerequisite materials, there is grave concern that the holes in the knowledge, poor academic habits, broken confidence, or a combination of all of these will likely lead to a lack of success in the first year of Algebra. So we have created a course for a subset of these students (limited by available resources and prioritized by what students are most likely to respond to an intensive intervention) that they take concurrent with their Geometry course in ninth grade called Reasoning Algebraically in Mathematics (RAMS). This RAMS course is offered to students and families as a way to plug holes in prior knowledge, preview material, and restore confidence in their mathematical potential. Giving up an elective spot can be tough for some students, but when compared to the potential of struggling and retaking math courses in the future, most recognize that losing one spot now to earn elective math credit is good insurance against failing a required course later that is part of a larger sequence that all has to be completed to graduate and will eat up elective spots while impacting their GPA.

There are a few critical points to make about the RAMS course. First, it is sold as a "leadership academy" of sorts; it is training them to be some of the best among their peers when they take Algebra together in the subsequent year. The goal is to take students who are not traditionally strong, and give them a year-long experience, with no pressure or pace, of understanding critical underpinnings of Algebra. We spend time delving into variables and relationships, and sort out independent-dependent distinctions; we talk about how quantitative variables that are related can be examined by listing pairs that go together on a table, or seen visually on a graph, or the calculation procedure can be recorded as a rule. We have time to play with interpreting graphs, patterns in tables, or what it means to "undo" a rule. We find math in the real world, and come up with strategies for "mathematizing" the functions we find. We practice explanations and justifications, and they get used to doing mathematics in a way that will benefit them moving forward. It is critical that the course is not seen as a remedial course for "dumb kids" to get caught up. Careful attention is paid to talk about the opportunity to get a preview of math so that next year there will be a handful of kids that already "get it" in the classes. Second, this course is taken concurrent with their Geometry course because we do not want to take them out of sequence with their peers or create a cohort of students moving through that have been "tracked." And third, this course is an opportunity to appreciate the process of mathematics, so conjecturing and wondering and trying ideas is honored as the way to best support nontraditional students. All too often (even in other buildings in our own district), math "intervention" means packaged modules "targeted" at content identified by traditional "screener" tests of basic skills—in short, the direct opposite of what makes mathematics interesting. In an effort to win hearts and minds back, we cannot double down on heartless and mindless.

So in ninth grade, all students are in Geometry, and a handful (set to be about 70 next year) are offered the opportunity to take RAMS as an elective. Then, *regardless of Geometry success*, all students are placed into Algebra A/B (the first year of Algebra) as sophomores. Again, they all have the option of choosing the regular or college-bound (Honors) version. A student who did not earn Geometry credit will need to take it at some point in the next few years prior to graduation (probably concurrent with the on-track math course), but in this way all of our 10th graders are together learning about function characteristics and solving with linear, quadratic, higher-order polynomials, and rational functions. Successful students then as juniors move on to Algebra C/D, which will include periodic, exponential, and logarithmic functions, probability, and statistics. As before, students are also welcome to choose the Honors version for college-intending students. Either way, at the end of their junior year, all students who do not re-take have the required State of Michigan content standards for graduation, and the prerequisite knowledge to take the AP courses as seniors (or, any of our non-AP options). The data from the RAMS course shows impressive outcomes: in 2015-2016, we saw that in a population where 100% were predicted to fail Algebra, 80% passed Algebra A and almost 50% passed Algebra B (content beyond what as previewed).

Closing Thoughts. In this section, we have talked mostly about student outcomes in the context of evaluating whether some decisions we made as a department are helping more kids learn more math. Usually though, student outcomes are judged by their own merits, to figure out how much mathematics a certain student understands for the purpose of evaluating their level of proficiency. In doing so, the evaluation is only as good as the assessment tool, and we have laid out some important commitments we have so that what we value is truly being measured. Perhaps the biggest potential downfall in trying to make significant change is to not have the tests match the goals of the course. Students will orient how they spend time and energy to being successful on the end of unit assessments, and teachers will work to prepare their students for success on them. If the tests look like surface-level applications of mathematical magic tricks (even sometimes with the formulas given to them!), then that is the ceiling you are likely to get.

So what are some indicators of a test that will push students beyond the traditional required regurgitation? It is not the absence of procedural questions, because we do indeed want to know if students can use mathematics to answer important questions. And, it is not simply what appears on the paper at the end of the unit either; a "good" test can be undermined by reducing the unit-long preparation into training students to answer those kinds of questions only. The assessment has to be considered (whether in practice it is or is not) a subset of all the possible things one *could* thoughtfully respond to if you had sufficient understanding from the unit.

That being said, there are a handful of criteria that might be reliable requirements of assessments that are true checks for understanding. The list may not be exhaustive, and some points can even be arguable. That conversation might be a really good starting place for a department to begin their journey of reform. But if I [Craig] were to make a list, I would say that a quality assessment would:

- Include writing. Students should be using language to express ideas, explain concepts, or justify reasoning. If we want to know if students understand the <u>why</u> and not just the <u>what</u> and <u>how</u>, then we need to ask.

- Incorporate Big Ideas. Students need to attend to important concepts that underlie the body of work that was just studied, and see the big picture.

- Have opportunities to demonstrate an ability to perform the actual math. We cannot determine who knows what and how much each student knows simply by counting how many circles they bubbled in correctly. If a student's understanding is being judged via right/ wrong dichotomies, and ones that could be random variation through guessing at that, then the conclusions are spurious and detrimental.

- Take advantage of questions that are written in a way that can check for misconceptions. This is not to say, however, that questions are written to trap or mislead students. The specific coefficients, wordings, or given information chosen in a particular problem are not arbitrary and choices must be made to check to see if students can handle variations. Take, for example, the case of solving polynomials: a test should determine what students understand about this concept by having some problems that have no solutions, some that have 1 or 2, and some that no algebraic technique can solve (i.e. $ax^5 + bx^2 + cx + d = e$). Or, if we have been taking advantage of the fact that zero divided by any non-zero amount is zero in order to solve for when two rational functions are equal, then we may want to check to see if a student will solve $f(x)/g(x) = 9$ by looking instead where $f(x) = 9$.

- Ferret out robots. An assessment of understanding needs to distinguish between kids who memorize, and those who are flexible.
- Requires professional judgement to score. This may very well serve as a litmus test—that an assessment needs professional expertise and sophistication to analyze and accurately determine the level of understanding demonstrated.

YOUR TURN: RADICAL CONVERSATIONS

This Radical Conversation follows directly on the conversation about curriculum from Chapters 3 and 4. With those conversations in mind, and perhaps a written summary of the key takeaways from that conversation, pull some student achievement data from the key time points you identified in which an approach to curriculum changed. Compare the ways in which the curricular changes you have made interact with student learning outcomes. At each of the time points, address two key questions: What did we **expect** to change with respect to student learning with this curricular change? and What **actually** changed with respect to student learning?

You will need to take care in this conversation not to attribute changes, or lack of change, in student learning outcomes to external factors like population shifts or changes in state assessments. While external factors can undoubtedly influence outcomes, teaching is a complex system and both internal and external factors come into play. Focus on the factors that you can change.

To close this discussion, identify aspects of student performance that you are not currently capturing that you would like to capture. This might include ideas local to the classroom, such as conceptual understanding of mathematics, or more global data-driven sources, like changes in course-taking trends over time.

INVESTIGATION AND REFLECTION ACTIVITY 5

Using the aspects of student performance that you identified with your colleagues at the end of the discussion, reflect on the action steps you would need to take to collect those data and track them over time. For example, if you wish to develop richer student assessments, who would you need to get buy-in from in your school or district? What would the development process look like? How might you determine from year to year if assessment items remained on the assessment or needed to change? What structural systems would you need to put in place to support the long-term storage and analysis of new data? And how might you go about convincing administration that the time and effort are worthwhile?

Next, reflect on the structures and partnerships that your district has, and write about the ways in which the district might develop future opportunities for stronger teacher collaboration, professional development, and outside partnerships. This might include internal ideas such as protecting common planning time, focusing faculty time on mathematics and student learning rather than procedures, or establishing a journal club for faculty to regularly engage with new educational research. Alternatively, this might include more external ideas such as pursuing a more regular partnership for student teachers with a local university, providing faculty with systematic opportunities for external professional development (through a state conference or national NCTM meeting), or identifying professional development providers that you would like to bring into the district to work with you and your colleagues. For each idea that you identify, make a clear plan for the person with whom you will follow up about implementing the idea, and on what timeline.

CHAPTER 6

THE ANATOMY OF A SHARED VISION

Beliefs About Mathematics Teaching and Learning

Take the set of all people that have been to school in the United States, and experienced math instruction in elementary, secondary, and often even postsecondary schooling. There exists a subset that have elected to go into the field of math education, and then again a subset of that group who have gone into math education programs that are designed to confront the assumptions about what it means to know and do mathematics and challenge previously-held convictions about what a math classroom ought to look like. And within that set is a final subset of math educators that have been able to maintain this vision, bolstered by the National Council of Teachers of Mathematics [NCTM], Association of Mathematics Teachers Educators [AMTE], National Science Foundation [NSF], Psychology in Mathematics Education [PME], TODOS: Mathematics for All, and other professional organizations and like-minded colleagues, while working in a system with incredible inertia to resist change. This final subset has a really small cardinality and is probably the biggest reason that, even with the technology that has been invented, math instruction across the country is fundamentally similar to what it was a century ago. Even in the most technology-rich flipped classroom settings, students may each be provided

A Quiet Revolution:
One District's Story of Radical Curricular Change in High School Mathematics, pp. 115–138
Copyright © 2018 by Information Age Publishing
All rights of reproduction in any form reserved.

with a laptop or tablet, but teachers still record their lecture and examples for kids to watch on them, and then use the "flipped" time in class to still help them with their homework by doing most of the thinking for them. That final subset consisting of the small group of radicalized math teachers are outliers among the spread of math classrooms, and as such can face overwhelming pressure to revert to existing within the interquartile range.

I [Craig] have often wondered what anchors this incredible inertia, given the obvious fact that American society as a whole hates mathematics so much that it is the perpetual go-to joke whenever education is mentioned. Why is everyone holding on to (and in some cases, running swiftly towards) a vision of math education that clearly does not inspire, does not work for the vast majority, and is contrary to virtually all the research on math learning that values understanding and being able to actually use it to answer genuine questions? You would think as soon as even a handful of people arrived offering something new and different, let alone effective, they would be greeted as liberators (to steal a phrase that was equally accurate). Instead, we see tremendous pressure put on them, often by people with little or no background in math education besides being a consumer of the education system as a student, to default to the widespread perception and comfort of the "normal" traditional math experience. It clearly is not nostalgia, and cannot simply explained away as fairness about a desire to subject future generations to a rite of passage that they themselves had to survive. Indeed, there is a self-righteous passion about it that implies that teachers avoiding the traditional "I Do/You Mimic" model are hurting children, and must be reined in. The backlash is well documented, and has played out in waves over the past five decades, most recently appearing as passive-aggressive Facebook memes mocking the antitraditional nature of Common Core Mathematics. Where then does this staunch resistance to change come from, and why has it been so effective in suppressing advancements in effective mathematics instruction?

I think there are two reasons to consider here that collude to create the inertia for things to stay the way they have always been. The first is that the traditional model has served a small group well, in that it gave them the cultural capital needed to play the game. And for them, their self-defined success is proof enough that that kind of learning is just fine, thank you very much. So even though their understanding of Algebra is little more than symbolic isolation-of-x tricks, that was enough for them to get a good enough SAT score to get into college and their memories were sufficient to reproduce the professors' work on similar problems for their undergrad math liberal arts requirement. Even without knowing what they do not know, their perception is that they were good at math; and these people with a positive recollection of that time are the ones who made it through college and are over-represented in the pool of active and well-connected parents. In other words, those that believe math instruction should continue to be rote and traditional are the ones that did well under those conditions, have therefore succeeded in schooling and enjoyed the experience, and make up the small contingent of adults who are active parents, educators, or School Board members. Those that would rally against a failed program are not often in the position to do so.

Second, requiring that students actually know and understand what is going on in mathematics requires more from them both in time and mental energy. People like me, who could derive any polynomial or put any matrix into row-echelon form quickly, had to spend very little time during class, and almost none outside of class, thinking. And, if a nontraditional pedagogical approach means that students must reason and think rather than just listen—rather, create mathematics and evaluate statements made by classmates—and cannot survive simply by memorizing what has been demonstrated, it requires some brain "sweat equity." Simply put, it is harder. It requires wrinkling your brow, and on occasion considerable time playing around with an idea. More time, more effort, and on top of it all, introduces self-doubt as an unfamiliar component to learning. And, it goes against the grain. Not exactly a recipe for comfort, and when you combine rigor with discomfort, it is fair to say that this new cocktail is hard for students. So, it is also fair I think to assume that teenagers are not often in a place to appreciate someone who is making something they need to do harder. And, there are many parents and guardians as well who (to be fair, out of love) want to save them from the potential emotional destruction of being in a hard class. And, there are many professionals in the building whose jobs and definitions of self-worth are measured by "getting their kids through." Having a hard class, then, is in conflict with counselors, at-risk coordinators, and/or special education case managers who see rigor as a barrier to their student reaching their finish line successfully. From their perspective, they are not helping a needy group of students avoid the kind of learning that may empower them to make smarter choices or avoid getting taken advantage of in the future; they are supporting students to be successful, as defined by checking off graduation requirement. As a co-worker in this position put on social media recently, "To me it did not matter how they [students on her caseload] did it, what mattered was they did it! The education system needs to be creative to meet the need of ALL students or the drop out rates will rise."

So the tremendous pressure of the system to remain a traditional mathematics program exists and can come in some form almost daily. Parents can send e-mails questioning the legitimacy of what you are doing, with no idea what you are doing. Comparing the amount of work that is required to prepare and assess students' thinking to someone else who just follows a textbook. Coworkers or administrators may secretly try to find workarounds to help students avoid the classes with creative scheduling or via mindless online programs. Students who would rather be told exactly what to do and how will complain publicly and privately. Adults who are not convinced with professional articles or specialized expertise in the area will talk in social circles and strategize their pushback. Without a confident stance, evidence at the ready, and a very careful balancing act of strong yet inviting public relations, teachers in isolation can easily be retrofitted with "I do, you do, we do" ideologies. They become prisoners of war, forced to teach in ways that they do not believe in but that are on the surface more understandable to the outside observer. But being understandable offers no honor.

If capitulation is not an option, communication is crucial to maintaining the vision of secondary mathematics education (at least until a critical mass is reached). This chapter attempts to talk about the deeply held beliefs about math teaching that exist, and how to work on bringing people on board to a new, shared vision. Oh, and spoiler alert—telling does not work.

SELLING TO STUDENTS

I am beginning with how to get students to buy in not because it is the easiest (quite the contrary) but because without it, none of the rest really matters. The people you spend your time with day in and day out need to be at least somewhat on board, otherwise it is like a coach that has "lost the team" as they say. This piece is, I truly believe, the difference between things working and things not, even though you swear you set everything up the same. So let me repeat the caveat here that is a recurring theme of the book—we do not have the answer, and what I have tried works some times and not others. My goal is to only share the struggle and offer what seems to hold the most promise, in the hopes that more accomplished folks than I will hit on the key pieces that raise the probability that more students will learn more mathematics.

One thing that feels really important is setting expectations on the first day. Students walk in with a clean slate and wide eyes, wondering what this part of the day will look and feel like for the next 90 or 180 days. This is not to say that they walk in with no previously-held beliefs about what learning mathematics looks like, what mathematics itself is, and who they are as learners of mathematics. But, at least on the surface, they are excited about what may be. Therefore, it seems like an important first step is to calibrate them to the notion of what they can expect to be doing. Let them know on the first day that class may feel different, because they will be asked to up their game now to doing mathematics instead of just hearing about it. There is a nice opportunity to contrast an unknown future with the arithmetic work of the past—rather than be quick and have memorized the times tables through 12 (you know, because integers to 12 are critical, whereas 13 and above is only for parlor tricks), we are going to study relationships in the world around us. We honor mistakes and thinking and wondering. We are going to work really hard, and so on.

It's important to understand that setting this up does not necessarily mean anything to them at this point. They can hear it but will not know what it is like to experience it, and they certainly will not know if they'll like it even if it sounds good at the onset. But the piece that seems to be important is that they know that whatever it is, it is purposeful. It is intentional, and comes from a place of knowledge and love for their growth. And as soon as introductory remarks and other minutiae get out of the way, we jump right in to do something that will demonstrate what class will feel like. It may be a short introduction to the first topic, or it may be something like a number talk (Humphreys & Parker, 2015) that will serve as an indication of how they will spend their time in this class. Whatever this initial task is, it needs to require very little traditional knowledge (or in current educational parlance, it should have many inroads for students). And as soon as possible, that thing we do should involve noticing, wondering, generating questions, working together, and somehow getting everyone to state something aloud for consideration. It's imperative that students define this space as somewhere where they will speak and think.

And the first week, students play along just fine. It is the first moments when that starts to break down where important reactions can make a huge difference. For example, I know for me as recently as a few years ago, I would pose questions and let students who happened to have a good idea share their thinking. By week two or three, students would start to settle into roles and issues of status began to get defined (for more about what we mean by status and how it plays out in the mathematics classroom, see Cohen & Lotan, 2014; Featherstone et al., 2011). Variables like impulse control, self-confidence, introversion, politeness, assertiveness, and even seat location determined the direction of the class, and whatever norms I worked to set up unknowingly started to break down. And after a month or more, I wondered why I was only hearing a few voices and lamented about the sheer number of kids "watching" class when I was sure my analogy in the first week about watching friends work out and expecting to get in shape had hit its mark.

This is but one example of how easy it is for the norms to break down via a missed opportunity for the teacher to notice and address. Often, early-career teachers make the mistake of thinking that norms are something to set and forget. While the setting of norms at the start is critical, norms have to constantly be monitored, checked, adjusted, oriented, and even renegotiated. But the trick is that each term or each course has different dynamics and a different way to fall apart that often does not make itself known until the symptoms become apparent and the damage is done. Teaching is hard, remember? And so besides sprinkling in work about the benefits of a growth mindset (Dweck, 2006) and making deliberate and explicit statements consistently about the nature of math and learning math, you have to organize your curriculum around student questions, teach through scenarios that may convince them that knowing math is worthwhile, avoid voluntary participation and isolationism, orchestrate group work, access prior knowledge, formatively assess, have personal relationships with each student, and do not let your guard down for even one second because it can all unravel with one or two bad decisions in the moment.

So yes, enacting near-perfect day-in, day-out in teaching is like catching lightning in the bottle. Luckily for all of us, that high aim is not an all-or-nothing wager. Every day we work hard to reinforce with students the right things, and over time these things do indeed seep in. The task is not hopeless, and what we do with students can impact their perceptions. Can we seriously undercut our efforts with a poorly designed or thought out plan? Yes—consider the case of trying to play both sides by having kids explore some days and other days be more direct with your teaching. Why would they wonder and explore when they know they can wait a day or two and get the answers from you? But absent such blatant missteps, we can make a difference on students' malleable perspectives about learning.

Two years ago, I asked students to take a survey on the first day of school that asked them to agree or disagree with a set of thirteen statements. I did it at the time because as a department we had bought into the work Jo Boaler was doing with Carol Dweck on growth mindsets, and I was simply curious to see if I could change the way students thought over a semester (18 weeks) by being attentive to the way I talk about learning ("Smart is not something you are, it's something you become through hard work"). I also threw in some statements about the nature of learning mathematics, just to see if existing in a nontraditional classroom impacted their thoughts on that as well. The statements were:

- Some people are born to be good at math and some aren't.
- There is hardly anything to argue about in mathematics.
- I see where Algebra is used in everyday life.
- Speed is a requirement for success in math.
- I'm hoping I don't make mistakes this year.
- I have been successful in math so far.
- Learning math is easy for me.
- If I work hard enough, I can be successful at Algebra.
- Knowing math makes my life better.

- Math may play a large role in my future career.
- I don't want to be challenged in this class.
- I work outside of school on my work.
- I am comfortable sharing my ideas in math class.

Arguably, these could have been better written or I could have included lots of other interesting ones; at the time, I was not thinking a wider audience outside of me or maybe the department would see them. Several of them (notably the first, fifth, eighth, and eleventh) speak directly to having a growth mindset, and the rest were just things I was curious about at that time. For each statement, I had students individually select a score from 1–4 with 1 being "strongly disagree" to 4 being "strongly agree." Some students wrote in 2.5 on some and I used that in the averaging.

In total, 78 students in the first year of Algebra took the survey (that included some students who were retaking) and resulted in the following incoming averages. I considered 2.3–2.7 as a neutral response (no shading), less than 2.3 as a general disagreement (dark shading), and more than 2.7 as a general agreement (light shading). Figures 6.1 and 6.2 reflect those three broad categories as a coarse visual tool.

Lots can be inferred from the results, and much more analysis would be necessary to be able to make any official claims. However, when I saw how the numbers looked, I was surprised to see that by far, the strongest agreement was to the statement that they believed they could learn Algebra if they tried. I had assumed that they would come in having written off math as just something many saw themselves as just not good at. Just as interesting, though, was that the only other score to average an "agree" (3) was hoping they did no make mistakes—a central tenet of having a growth mindset is that making mistakes helps you learn and is not evidence that you are not a "math person." I also noticed that the middle grades, having

Some people are born to be good at math and some aren't	There is hardly anything to argue about in math	I see where Algebra is used in everyday life	Speed is a requirement for success in math	I'm hoping I don't make mistakes this year	I have been successful in math so far	Learning math is easy for me	If I work hard enough, I can be successful at Algebra	Knowing math makes my life better	Math may play a large role in my future career	I don't want to be challenged in this class	I work outside of school on my work	I am comfortable sharing my ideas
2.58	2.24	2.37	1.98	3.06	2.31	1.90	3.38	2.64	2.71	2.47	2.85	2.37

Figure 6.1. Start-of-term mindset data from Craig's Algebra classes.

switched to Connected Math Project starting in 2008–2009, seemed to
have some of the desired effect: students apparently see mathematics are
something to debate and talk about instead of clean and correct statements
of fact.

At the end of the term, I gave them the same document and had them
rank their thoughts again individually.

Some people are born to be good at math and some aren't	There is hardly anything to argue about in math	I see where Algebra is used in everyday life	Speed is a requirement for success in math	I'm hoping I don't make mistakes this year	I have been successful in math so far	Learning math is easy for me	If I work hard enough, I can be successful at Algebra	Knowing math makes my life better	Math may play a large role in my future career	I don't want to be challenged in this class	I work outside of school on my work	I am comfortable sharing my ideas
1.96	2.05	2.60	1.83	2.65	2.48	2.19	3.46	2.64	2.72	2.40	2.55	2.68
*												*
-0.62	-0.19	+0.23	-.15	-.41	+.17	+.29	+.08	.00	+.01	-.07	-.3	+.31

Figure 6.2. End-of-term mindset data from Craig's Algebra classes.

I do not know what kind of change we might expect to see (the change
of one from a 2 to a 3 is a major shift from agreeing to disagreeing, which
is not at all similar to a change of one from a 3 to a 4), and I do not know
if one term is enough to really see the full impact. I will leave most of the
interpretation to the reader to determine how much change is interesting
for what intervals or what any of this may or may not mean, but the point
is that students do change their perceptions as a result of the messages they
receive from the teacher, and we can move large groups of students into
more productive mindsets.

When people ask me what I do, I usually do not tell them that I teach
mathematics. Instead, I like to reply that I create experiences for students
to learn mathematics, and smile with a self-congratulatory nod for being
so clever. But truth be told, what I really should say is that I attempt to
get students to buy in to learning math, because that is really what it is all
about. If they cannot suspend disbelief for long enough to take a chance
and feel what it's like to play along, then they are walling themselves off
from real, genuine learning. At its heart then, my job has to be to get them
to tear down those walls ... or maybe peek over them for a bit and then
chisel a small hole here and there. This can get oversimplified by describing it as teachers just wanting students to try, because I want them to try
the right way. Students can put forth effort the way I would like them to
(having curiosity, playing around with an idea, asking questions, talking
aloud when working together in small groups, articulating dilemmas that

have arisen in their thinking, etc.), or in the way that I do not need (writing down or working to memorize someone else's thinking, getting a tutor or using an Internet search to circumvent the explorations, doing and redoing procedures from class at home, reducing complex ideas into misleading rules of thumb, etc.); and without knowing any better students will default to the latter, thinking they are doing what they are supposed to and likely end up worse off. Additionally, that statement of just wanting students to try negates the teacher's role in crafting experiences that hope to compel them to—over and over and over again.

SELLING TO PARENTS

The challenge of communicating to parents and guardians is that their experiences in school, or more likely their memories of their experiences in school, are what informs their response as much or more than what the teacher may actually say. And unlike most of the students, the sheer range of those experiences in age and variance in geography can mean an incredible difference from one family to the next. One may have extremely negative memories from their high school experience, even from a world away, and come in with a deeply personal and emotional mistrust of the school system. Another may have had a bad run-in with your principal from an older child and be predisposed to being argumentative. Another may be one of those that had a supertraditional math classroom back in the day and loved every minute of it. One may trust you, and another may not; neither with any good reason. One may see themselves as a "math person" and be frustrated that they cannot seem to understand what their kid is doing. One may get their politics from websites that push them to speak out against the Common Core. Others may be uncomfortable to see students using a calculator, having too much homework, or having too little.

And, with few exceptions, the adults you interact with have not been steeped for years in the advancements in brain research or teaching mathematics. Their love for their kid that is having trouble will trump your résumé just about every single time. This is not to say that the vast majority of parents are problematic; in fact, it is quite the opposite. But with upwards of 150 sets of student parents in your life every year, the probability that someone will question something is greater than zero, and it is raised higher when you are teaching in a way unfamiliar to them. And in days of social media and involved parents, it just takes one or two to talk and it can become something bigger. But even if it is just one parent, for a student that may otherwise be on the fence or up for deciding if they want to trust you and play along, an undercutting comment at home can carry more power and influence than months of work with that student in class. Add in the fact that a parent does not get to experience the classroom, hear the directions, notice the careful way you attend to different needs, see the effort to include all kids, or listen to the way you interact with groups—they largely get their information through the lens of a teenager who has a vested interest in having the parent see things their way—and you can see how communicating a shared vision can be critical to success.

For starters, I begin the year with an e-mail message to all the guardians of the students assigned to my classes. This idea of introducing yourself to families and sharing your hopes for the school year is a tried and true one; but I speak specifically about a couple things that experience has taught me can be instigators of issues later.

WHAT CAN I DO AT HOME? A HELPFUL GUIDE FOR PARENTS OF MATH STUDENTS

"I don't see them doing much work once they get home..."

"This math is beyond what I learned or can remember..."

"Without a textbook, I don't know how to help at home..."

These are common and valid concerns for parents or guardians of students taking a high school math course. Luckily, all of these are addressed through our curriculum design, albeit what students do at home may or may not look the same as what we remember from our high school days. For one, it is true that both the level and the amount of mathematics required is far more than what it took to graduate in the past. Second, much of what we now know about how people actually make sense of mathematics has evolved; and therefore, there is much more group work, inquiry, and using technology than in the past ... all of which are difficult to recreate outside of the classroom experience.

So what we see students doing at home may or may not be content that is familiar, and even if it is, it may not be understood in the same way that you recognize. And, while there are some times where they will have work at home to practice ideas or strategies that are already in place, other times they may be asked to consider or think about new questions and come up with ideas that make sense to them at that moment.

(E-mail message continues on next page)

SO WHAT CAN YOU DO AS A PARENT AT HOME TO HELP UNDER THIS NEW PARADIGM OF MATH EDUCATION?

First, it is important to know that all students are keeping their notes, ideas, and day-to-day work organized in what we call their "unit packet." In essence, this is their textbook—it is where they go when they need to look back at previous examples or explanations, it's where they are given work to do; and it is a record of their learning and completed assignments. Each unit that we do will be a chapter in the year-long book we are creating about the study of the course content. Ideally, it is both theirs to keep for future use and written in their own words, which should make more sense and be more meaningful. It has a table of contents with all the worksheets and important papers listed, followed by those assignments in that order, interspersed with their notes as they go. Until the end of the unit, when this packet is submitted for credit, we consider it a "work-in-progress" that is continually added to and adjusted as we encounter new learning related to the main ideas of the unit. Therefore, if they have a math question at home about what we are doing in class, they should have their unit packet as a useful resource.

Second, it is also important to know as a parent, you do not need to be responsible for being a mathematics content expert at home. Besides the previous work we have done and gone over in class that is recorded in their packet, we are also available for extra help outside of school hours to fill in any holes your student may have or any idea they may need extra help on. Understandably, transportation and extracurricular activities can sometimes conflict with this opportunity, but in our experience, just a little bit of time outside class once in a while pays large dividends. Besides that, the biggest thing you can do would be to encourage your student to ask questions in class. We set up the learning environment based around questions and ideas, so a student that is quiet and just "watches the show" will end up just trying to memorize other people's ideas. We often tell them that listening to a music teacher explain how to play an instrument or a coach tell how to play a sport is not enough—you need to do it yourself. Math is no different in this respect.

BUT IF I'M NOT SHOWING THEM WHAT TO DO OR HOW TO SOLVE PROBLEMS, WHAT SHOULD I BE DOING?

1. You can expect that they will have some work to do at home at least a couple times a week:

   - In the times when they have practice problems to do at home, the intent is that they know enough mathematics to answer the questions. If they get stuck, you can encourage them to look for examples in their packet, and then make a plan to get extra help.

   - If the homework is comprised of questions that are asking them to consider new ideas or try and make sense of something before the ideas have been solidified, then the intent is that they just record their thoughts and what makes the most sense to them at that moment. This will serve as a starting point for discussion to generate new mathematical concepts, and the thinking they do will be crucial to their ability to engage in the important thinking during that conversation. Be sure that they explain their thinking and, if they get stuck, have them write their question(s) down instead. Just be sure that they spent some time thinking about it and come to school with something thoughtful written down.

(E-mail message continues on next page)

2. If they do not officially have homework, you can ask them to bring their unit packet home. You may want to do this periodically anyway if they are not bringing it home often enough. Make them sit down at the table and organize it:

- the table of contents should be filled out up through the most recent assignment,

- the papers should be in order and organized so they can find information when needed, and

- every previous assignment should be fully completed.

3. Thumb through the previous handouts with them, and identify any incomplete answers for them to finish. Look especially for places where they may have written down what was on the board without any work, or put an answer down without responding to the "how" or "why" part of the prompt. Start at the beginning of the current unit and go through question by question and worksheet by worksheet until everything is fully complete.

4. If everything is already complete and organized in their unit packet, the single best thing you can have them do is choose problems from previous assignments and cover up the answer that we had discussed in class. Then, treating it like a question on a test, have them think about it and write out a good response. When he or she is finished, you will be able to compare his or her answer to the one on the worksheet. We find that, more than anything else, this strategy is best for ensuring they understand the ideas as we go, and the most effective studying for assessments.

Most importantly, thank you for your support at home. Anything we can do together to make sure our students know that we are all on the same team to support them in their learning will have a huge payoff in the end!

While this letter is specified for the way I organize my classroom, take note of a couple of important things that are universally useful. First, making an explicit statement that the parents and teacher are on the same side defines the relationship as sharing a goal of helping the student learn. It may end up being true that the teacher and some parents differ on how best that happens, but knowing that all parties involved have the same interest at heart from the get-go sets the stage for working together. Secondly, I honor the concern of a parent trying to adapt to the new way of math teaching rather than disparage it. Rather than make them feel defensive about how they feel or get into a disagreement about how it ought to be done, I prefer to let them know how it is going down (because, to be fair, we are not on equal footing) and then be helpful about what they can do to exist in this new world and still support their kiddo. Third, it also lets them know right away that things will be different, and that we have taken pains to still set them up to be successful. In the long run, that's all parents want—a fair chance for their kid to learn in a classroom run by someone who knows what they are doing.

Periodically, I will also send e-mail updates to families as we go. Sometimes they are under the auspices of keeping them abreast of coming timelines, but regardless the intent is that they are comforted that I am being thoughtful about where we are at and what we are doing. Parents who may know that their own understanding of what's happening is not the large picture still are consoled when they can tell that the teacher is working hard and thinking about what kids need and how to get them there.

Unfortunately, the other piece that seems to factor in is the age and experience of the teacher. I am no doubt much more articulate about my choices as a teacher now, but I also feel that I am given much more automatic credibility now by parents than when I was a teacher in to my first 5 or 6 years. By nature, parents may naturally question someone new to the profession (which is understandable); but new teachers frequently do not come across as confident as more experienced ones do, which could seem like blood in the water to a questioning parent. This is not to say that experienced teachers that have been in the district for a few years get a free pass—if a narrative about them, right or wrong, reaches a large enough set of people, then students will have preconceived notions that may never be overcome.

Besides sending a few measly letters, then, what can be done? While you can never guarantee to prevent issues, you have to be able to respond promptly and articulately when they start to creep up. I often will offer to meet with a parent or guardian that contacts me wondering about some things that I do in my classroom right away, so we can talk eye to eye and they can see I am a human doing what I think is best and not some mythical teacher that hates kids and sits back with his feet up on the desk while students flail aimlessly. I make sure I can justify all the decisions I make, and share the rationale for what I tried. I do not make statements I cannot back up, and I do not do anything in class with students that I would not want a parent or administrator to see. And while I am almost never taken up on it, I have a copy of *Principles to Actions* that I can lend to parents who would like to know more. And finally, I stay active in the math education community: not just to stay up to date on things and continue to be pushed (although that's the primary reason), but also because it helps me have practice talking about my practice and connections to those who talk about theirs.

SELLING TO COWORKERS

The challenge for helping coworkers understand what you or your department is doing is that most all of them fit squarely in the subset of adults who were successful and enjoyed school, and may be inherently resistant to something that is not what they experienced. And as mentioned earlier, some of them are focused on the most expedient way to get students to graduation, and not necessarily on how much struggle and time it may take to get someone to the level of understanding we would want for a citizen and member of society. And, the time students have in school is a zero-sum game: you take more time for a student to learn mathematics, and that means less time for them to have to sit in other classes, which teachers of those subjects see as every bit as important (and usually more) than Algebra. Often, advocating for a support class for students for all the right reasons is seen as taking students from elective programs, which is the lifeblood of those teaching positions. When the value of their jobs is on the line, it is easy to understand why some colleagues can feel a core vs. elective us-or-them mentality.

Much of the communication strategies for student or parents are in play here—notably crafting a personal relationship, having a thoughtful rationale, and being able to articulate your thinking while being a fair listener. The advantage you have with coworkers is that they see how hard you work and advocate for students on a daily basis. Whether or not they agree with everything you may say, they at least will respect where you are coming from—an advantage that is not a given with parents or students.

The other good news is that they trust and listen to professional work—educators are much more likely to be swayed by professional literature or impressed by work you do as part of a larger math education community. You can talk about your social constructivist approach (see Chapter 1 for a simple narrative if you need one) and it sounds like an informed colleague and not a jargon-spouting elitist. When people you work with continue to hear about things like presentations, publications, or awards that members of your department receive, they understand what it takes to be involved in that world. Credibility matters. When I was pushing an assistant principal several years ago about an idea to allow students to take courses at the local community college while still in high school instead of the senior electives we had, my main concern was that the college course was the epitome of a traditional math class—a professor, likely not trained in mathematics pedagogy, would be lecturing about a disconnected laundry list of procedural topics. After working for years to avoid that for our students, I did not want to all of a sudden increase the availability voluntarily. When the administrator countered with, "How do we know what they are doing isn't better? That we shouldn't be doing more of that?" I decided to stop declining an invitation to be nominated to apply for the Presidential Award for Excellence in Math and Science Teaching that someone from MSU had been suggesting. It requires a pretty exhaustive audit of your teaching, from writing up your philosophy to video vignettes. Most importantly, it is evaluated by an outside group of experts. And when the following year they identified my teaching as one of five finalists (for which the National Science Foundation does a great job of publicizing on your behalf), I had my answer to that administrator: that's one way we know what we are doing here is worth fighting for. More importantly, it was a reaffirmation to stay with my gut, my training, and my experience in times when self-doubt or exhaustion can creep in. Having something published, presenting at a state or national conference, or doing work with university-based teacher education programs means something to coworkers, and serves a dual purpose: making you better at your job, and building experiences that can help you advertise what you do with colleagues.

Building trust with adults that work in the building and earning external credibility means you can talk openly and honestly to them. For example, you can reason that students who defer an elective to take a support class as freshmen may be able to be successful in the Algebra sequence and ultimately have more room for electives as they go through school, and have the statement stand on its merit rather than dismissed out of hand. It also gives you a place to reach out from in sharing articles with them. Recently, there has been a spate of articles written for a general audience (and often shared on social media) that have described the myth of the existence of a math gene (see Ferro, 2013; Kimball & Smith, 2013 as two recent examples). Throwing a copy in people's' school mailboxes with a quick, "thought this was interesting" signed note at the top can plant seeds among colleagues to have future conversations. Many will glance at it and recycle, but some will also read and consider. And just creating the awareness that there is something else going on in math education that they may not be fully aware of is a huge first step.

Thus simply making coworkers aware—whether they be counselors and case managers, administrators, or teachers in other departments—of the kinds of work you are doing to improve the teaching and learning of mathematics is a significant achievement. Gaining credibility through professional projects allows you to have a standing from which to educate them on what and how that vision is different than what they may assume. Being able to reference and share professional literature and research (Principles to Actions' executive summary [NCTM, 2014] at the very least) changes the game from a debate of beliefs and perspectives to a discussion based on what we know works best for effective and true learning of secondary mathematics. And with a basis of collaborative respect, finding opportunities to advocate for policies and structures to support the learning of mathematics in your school and/or carefully call out ill-informed assumptions or clichés can be possible.

Even with the best intentions, confronting outdated or discredited views of learning math with other educators can be subverted with circular logic right out of a Lewis Carroll dream: stereotypes about the existence of "math people" (of which you are, for people that hold that view) can pigeonhole you and serve to undermine your message about the very perspectives you are trying to confront. For instance, justifying why you cannot just lecture traditionally once in a while as a "middle ground" can easily be disregarded as coming from someone who "operates in black and white" and is too "left-brained" to be able to effectively see anything outside of "right or wrong" dichotomies, no matter how carefully one may describe the incompatible contradiction in asking kids to be the experts on one day and then have the teacher operate as the "sage on the stage" the next. Instead of hearing the problem with textbooks having an introductory Explore activity and then, two pages later, having what the kids were supposed to discover in a bright pink box, coworkers may just presume that as a math person, you are less comfortable with reading or just shun books in general. And thus, while not necessary, it becomes really helpful to seek out allies among coworkers that understand where you are coming from, and may be more impactful being an emissary than you might be. Skeptical coworkers hearing a similar argument from a nonmath teacher carries power that you may not be able to conjure.

I have found the best way to bring people along is to have them be in the classroom with you. Back when we had coteachers from the special education department, it was not uncommon for them to talk about how they "get it now" after teaching a math class together and seeing it in practice. One of them is one of our biggest and most respected advocates still today. A particular former Curriculum Director took just one or two visits to shift from advocating for direct instruction in mathematics to liking what he saw. (And interestingly, instead of discrediting his prior arguments, he instead chose to redefine what he termed direct instruction to include what he observed. But, OK. It did remove the pressure of someone in power second-guessing the work we were trying hard to sustain and improve on.) The specter of what they imagine a nontraditional classroom to look like is often akin to the teacher sitting at their desk while kids try to aimlessly explore their way into rediscovering differential equations. When they instead see an artfully choreographed task and subsequent conversation about the thinking that is happening, it feels much less ridiculous than they imagined. In our current era of ubiquitous video recording devices, you can also accomplish this by taking and sharing video of your teaching, effectively inviting your entire school staff into your work if they choose to watch. And if you are fortunate enough to have them observe or participate over time so that they can see the growth of conceptual understanding, they may be able to notice for themselves the outcomes that otherwise sound abstract and ethereal. When the connections and building up of big ideas can be experienced, they have a place to understand why you would criticize an approach that features a shopping list of topics to "cover." Finally, sharing with them the very real dilemmas that teaching this way elicits invites them into the conversation we really want to have and changes it from "what should a math classroom look like" to "what things can we do to make this classroom be even better?"

Outreach can also be a purposeful event with the intent to bolster a common understanding. Several years ago, our department thought it may be useful to use some of our district's professional development time to have our staff experience a simulated lesson of what it looks like in our math classrooms. We separated our staff into groups of about twenty, and had members of our department facilitate an experience where the groups would read a situation and be asked to think about how they would solve it, similar to how we orchestrate it with our own classrooms. We gave them some individual think time to play around with their thoughts, and then had them share ideas with two or three others to try and decide what they thought the answer was that made the most sense to them was at that time and why. We then collected their strategies and discussed possible solutions as a group. The goal was for our colleagues to have a common experience and a framework in mind so that they may be more aware of what we mean when we talk to them about our work, and maybe also be able to respond to questions or concerns if overheard. We wanted them to feel what it may be like to not have an initial idea but then still walk away with understanding through a whole-group conversation. Prior to that day we also asked them to read *Mathematics Inside the Black Box* by Jeremy Hodgen and Dylan Wiliam (2006), which is a succinct and convincing synopsis of some of the big ideas in math education written by an author most of them had affinity for already from early work with formative assessment. Coworkers would reference that PD day for years afterward.

INSTITUTIONAL REINFORCEMENT

The previous section was mostly about crafting a shared vision by trying to work with those who have not spent their career thinking hard about teaching mathematics; but this is not to ignore the importance of fostering a shared vision even within your department. When I came into Holt's department, and well into the time chronicled in *Embracing Reason* (Chazan, Callis, & Lehman, 2009), our department was much more fractured than it is now. NCTM (2000) had published their Standards and our department was by and large working toward that approach, but there were still several members who taught traditionally in their classroom and tended to sit back during department meetings. At times, conversations could get contentious. However, over time, staff turnover with retirements and replacements have gotten us to probably the most cohesive and "on the same page" as we have ever been. Having a consistent and unified message carries a lot of power.

Intentional hiring, therefore, can be a critical component. It has to be acknowledged that acquiring new colleagues in mathematics is not dissimilar than what a President does when selecting nominees for cabinet positions: assuming people that do not have the necessary experience and work ethic or sufficient background are weeded out, selection depends on their philosophical orientation. Just like a Democratic President would not likely nominate a Conservative to the Supreme Court, your mathematics hiring committee should look to discern a candidate's perspective on teaching math. It might seem overstated to put it in such stark terms, but when you have to work so hard to get students, parents, and even the school to buy in, you do not want another math teacher undermining the progress, even unintentionally.

Luckily, at Holt we have a few things that happen to be in place that support us in our attempts to do so. First, we have maintained a connection to MSU's teacher education program in mathematics, which among other things usually means hosting a couple of interns each year. Ideally, we get candidates trained in effectively teaching mathematics that we have been able to work with and further train within our own building, essentially having year-long job interviews. We do not have to guess if we are able to teach well in the ways we want. Second, we have a long-standing norm in our building that a committee of teachers is involved in the hiring. The building principal ultimately makes the official decision, but historically we have operated where 1–2 building administrators will sit with 6–8 department members, and will collectively interview and decide defaulting to the expertise of the teachers. In those interviews, we attempt to ask questions specifically written to reveal the candidates' commitments. Finally, in the prior 2 years, our district has worked to get representatives from each grade level in mathematics together and craft a district vision for mathematics. This vision is very much in line with what we are describing in this book, and while it is in its infancy, the process has given us grounding when hiring if there is a discrepancy among the group. In hiring for the coming school year, for example, there was a candidate that the math teachers had low on their list (due to responses that indicated that her version of a typical class was going over homework, lecturing on the new topic, and then walking around helping when they get stuck on problems; and that if a student was struggling, she could re-explain it in a better way, etc.) but was actively involved in her prior school, had experience, and threw out a lot of edu-jargon that the administrators loved ("data-driven" particularly). And while it was clear that she was a very conscientious professional that I as a building principal would probably love to have on staff, she had beliefs about teaching mathematics that would be counter to the direction and passion of the department. We were able to use the district vision statement to cut through the discussion and simply say that she does not fit and we need to move on. Having a litmus test that is semiofficial was more convincing than having opposing viewpoints around a table.

Implicit in this structure is the recognition that not everyone certified to teach math is essentially the same. Districts have lists of what employees are highly qualified in, and can make staffing decisions based on this. When resources are held back from State governments, it is judicious to try and tighten the purse strings by trying to not replace retirements and absorb the position somewhere as a more humane mini layoff model. Often logistically, this requires administration to assert their "right of assignment" and move people to different departments or even grade levels in other buildings. And while there is an emotional component to this, from their perspective they treat a "good" certified teacher as "good" for whatever they happen to teach. In an effort to increase efficiency, a district may reassign a middle-level science teacher to teach in their certified minor to fill out a short schedule in Spanish at the high school. And, in many cases, going through the Spanish curriculum for someone that is certified may be completely doable. But in mathematics, for a building that teaches non-traditionally, this is likely not the case. Even having a math minor that is aware of the idea for what we are trying to do, unless they have seen it in practice over a significant period of time, the nuances that make or break it are not likely to be in place. In fact, even for me having done nothing but thought really hard about the teaching and learning of math for my whole career with people supporting me, I still have not figured it out. To be fair, it is hard to pull it off, even if you get it in principle.

To that end, we come full circle back to having purposeful strategies in place to reinforce the shared vision for those that are already on board in theory and working together to teach math better. Collaborating is critical to pushing one another to continue to work on improving practice. Carving out specific times to meet as a department, seeking out colleagues to talk through questions, content connections, or approaches that seem to have led to more fruitful student ideas, and having a shared drive for people to share their assignments are among some important ways we have supported each other. These common experiences reinforce and reinvigorate the hard work, and a sustained energy and passion is a really effective public relations element.

YOUR TURN: RADICAL CONVERSATIONS

This Radical Conversation provides you with some interview and question stems with which you can begin to engage departmental colleagues, administrators, students, and parents in discussions about their beliefs about mathematics teaching and learning. This activity directly targets the ideas discussed in this chapter related to growth mindset, the role of speed in mathematics, and emerging research on the roles of productive struggle and authenticity in learning mathematics. Through these discussions, you will have the opportunity to challenge unproductive beliefs about the role of speed, procedural work, and "inherent" talent in mathematics and consider how stakeholders can fully embrace a more progressive mindset about how students learn mathematics.

Choose one of the constituencies shown in Table 6.1 that represent the different "Circles of Encumbrance" (plus students) we described in Chapter 1. Gather your group in a natural setting that has space for such conversations —we recommend this rather than calling a special meeting focused only on this topic as it is likely to increase comfort levels. You do not need to gather a large group to do this work—as few as 2–3 people may be productive in some contexts. If you are interested in investigating student perspectives, using the prompts across multiple class sections would be useful to generate a diversity of perspectives. (You should also make it clear that their responses will not have an impact on their grades, or provide opportunities for them to submit written feedback anonymously in addition to a discussion.) Use some or all of the questions and sentence starters as you see fit.

When you conduct the discussions/interviews, try not to write anything down if you can. Writing during a discussion can negatively impact participants' comfort levels and make them feel like their responses are being carefully monitored. You are not looking to do that in this case—just to get a feel for their perspectives. After you complete your discussion, write in your Radicalization Handbook the important points you took away from the conversation. For each point, try to identify a next step—a more focused conversation to have, a research finding you would like to share with this person or group, or a mathematical task or video clip of teaching that you would like to share.

Table 6.1.
Prompts for Inviting Conversation by Audience

Target Audience	Discussion Prompts
Students	When do you feel most successful in mathematics class?
	When do you feel least successful in mathematics class?
	What are three characteristics that you think are important to have to be successful in math?
	Complete the following sentence: I could be more successful in mathematics class if...
	Is being successful in mathematics more like being successful in sports or being successful in business? Why?
	If you could travel back in time and tell third-grade you one thing that you thought would help them be more successful in math, what would that be?
	Does effort matter in being successful in math? How?
	You might have heard people say that it's important to learn from your mistakes. Does that happen in math? If so, how? If not, why not?
	If there were no limits on the time and energy you could put in, could you be excellent at doing mathematics?
Your Department	What are three characteristics of students that you feel are most important for students to have to be successful in math?
	Prior knowledge and effort can both be important for student success. How do you see the balance of those two ideas playing out with students who have been successful?
	If you asked students to complete the following sentence, what would they say: I could be more successful in mathematics class if...
	If we had no constraints on time, energy, and effort, what are two things that you would change about how you/we teach mathematics?
	What does it mean in terms of our teaching to say "all students can learn?" What specifically do we do to live up to that statement? What could we be doing but are not *yet*?

(Table continues on next page)

**Table 6.1.
(Continued)**

Target Audience	Discussion Prompts
Other adults in the building	What were your experiences like as a mathematics learner? When did you feel successful? When did you struggle?
	What was the most rewarding about your learning experience and what did you find least rewarding?
	When you hear students talking about their experiences in mathematics class, what do they talk about (good, bad, and neutral)?
	How has teaching and learning changed in your content area since the start of your career? What are the new techniques and advances in your field that I should understand?
	What is an idea that you wish you had learned about in mathematics but did not, or an idea that you wished you had learned more strongly/better/more deeply?
Building/district administration	What were your experiences like as a mathematics learner? When did you feel successful? When did you struggle?
	When you hear students talking about their experiences in mathematics class, what do they talk about (good, bad, and neutral)?
	If you interviewed our students on their last day of high school and asked them what important things they had learned during their time here, what would you hope for them to say?
	We often say that all students can learn. What does this mean to you in the context of our mathematics coursework?
	How can or should we factor a teacher's disposition towards their content into a hiring decision?

(Table continues on next page)

Table 6.1.
(Continued)

Target Audience	Discussion Prompts
Parents & the community	What were your experiences like as a mathematics learner? When did you feel successful? When did you struggle?
	What does learning mathematics mean to you? When you think about a classroom of students engaged in learning mathematics, what do you picture?
	What are the most useful parts of mathematics to you in your current work?
	What would you like to better know or understand about your student's experiences in math class?
	What is an idea that you wish you had learned about in mathematics but did not, or an idea that you wished you had learned more strongly/better/more deeply?
	When you think about solving a complex problem as a part of your work, what sorts of skills do you use?
	When you look to get better at an aspect of your work, what sorts of things to do you? How do you know you are improving?

INVESTIGATION AND REFLECTION ACTIVITY 6

Design a professional learning experience or artifact that you would like to use to try to shift the beliefs within one of your Circles of Encumbrance. Use the outcomes of your Radical Conversation for this activity to decide a good entry point and a good group with which to work. If you are engaging your mathematics department, you might think about a research-based reading (e.g., growth mindset, formative assessment, the equity section of *Principles to Actions* [NCTM, 2014]) and a few discussion prompts to engage in meaningful conversation about that reading. If you are engaging other adults in the building, you might think about a model teaching experience that engages them in some of the mathematics teaching and learning that you are interested in fostering, or inviting them to observe a class that you are teaching with students and have them analyze learning. (The observation tools included in Chapter 7 may be useful here.) For administrators, you may design a short presentation making use of student learning data you collected (see the next chapter for some specific ideas) coupled with some questions about current beliefs and structures about mathematics teaching. For parents and the community, this might be your own version of the e-mail about what to expect shared earlier in the chapter, or a community mathematics night (see next chapter for some thoughts about how to structure such an event).

Just as you would a classroom lesson, identify some tangible goals for this learning experience. What will you hear or see from the group that will tell you that you are successful? How will you respond to challenging questions and dissenting views in constructive ways? How will you provide all your participants with opportunities to have voice, and to connect your work to their current understandings about mathematics teaching and learning?

Conduct your learning experience. Record what you learned, the extent to which you met your goals, and what you might do differently the next time. And then identify a colleague to be your accomplice for the next iteration that was not involved in the planning the first time. This is how the revolution grows.

CHAPTER 7

CREATING YOUR BLUEPRINT

Professional Learning Activities to Support Systemic Change

The final chapter of *A Quiet Revolution* brings together the ideas that you have explored in the previous chapters that document the story of curricular change at Holt High School. There is, of course, no way to provide you with a clear, unimpeachable set of steps to reproduce the work. And even if we could, such a procedural structure would deny you and your colleagues the rich, conceptual understanding that we seek in the teaching and learning of mathematics, would it not? All right then. So in the absence of such a blueprint, we present a series of activities, tools, and frameworks that will support you and your faculty in having the meaningful and often challenging conversations about what it means to teach and learn high school mathematics. These activities are designed to lay the groundwork for the sort of iterative change that has led to the evolution of Holt High School mathematics over time.

We organize these tools and frameworks into three sections. As we have noted elsewhere in this volume, establishing a shared set of values within a faculty is a requisite first step in deciding how and when to make changes to high school mathematics curriculum. Our first section presents activities and tasks designed to support you in discussing, developing, and refining a shared set of values around the teaching and learning of mathematics. Data about students' mathematical learning is often limited in the extent to

---

A Quiet Revolution:
One District's Story of Radical Curricular Change in High School Mathematics, pp. 139–175
Copyright © 2018 by Information Age Publishing

which it informs what we do in the classroom. As such, the second section provides tools and strategies for stronger analyses of the data that are available to us, and give suggestions for straightforward ways to collect and analyze more meaningful data about student learning. Finally, the broader community needs to have a modicum of understanding about the changes your faculty intend to make and how you intend to make them. The final section of activities provide outreach and advocacy solutions that can help you avoid parental (or worse, school board) blowback as you revolutionize mathematics teaching and learning in your district.

Although the three sections outlined here are relatively sequential at a large grain size, the activities themselves are not designed to be engaged in in a particular sequence. It is likely to be productive to move flexibly between the categories based on your needs and the needs of your faculty, and the changes you would like to make. Most of the activities can also be revisited and/or sustained over time as well.

PART 1: DISCUSSING, DEVELOPING, AND REFINING DEVELOPING SHARED VALUES

Activity 1.1: Values Continuum

The Goal

Identify the current set of values that are represented in your mathematics faculty with respect to mathematics teaching and learning.

The Task

Generate a series of statements that represent different stances on mathematics teaching and learning. Some examples are provided below that are focused particularly on equity. You could also create or adapt statements that represent different aspects of growth or fixed mindsets (Dweck, 2006), or the productive and unproductive beliefs about mathematics teaching and learning listed in Principles to Actions (NCTM, 2014).

Provide each teacher a copy of each statement, preferably on a small sheet of paper like an index card that can be moved around. Have teachers consider on their own where they would place each statement on a continuum of agree to disagree. You might give different sets of statements to different groups of teachers and assemble them all on a single continuum at the close of the activity.

After teachers have had an opportunity to consider each of the statements, ask them to collaborate with a small group and place them on a coconstructed continuum. In doing so, they must make an argument for what the statement might mean, why the statement belongs in the place that it does, and how the statement is embodied (or not embodied) by current practice in the school and district.

Assemble all the statements from each small group in a continuum in front of all teachers. Have them choose statements to discuss that they feel are core to the values of the department, and others that they think are less core to the values of the department or that they find their placement problematic. The language of noticing and wondering (as described by Smith & Stein, 2011 among others) can be helpful here as teachers are pressed to describe the ways in which the values are or are not evident in their practice. For example, you might model this by saying, "I notice that we believe that all students have the capacity to learn meaningful mathematics. I wonder whether our course structures and decision making are truly providing everyone access who wants it." This keeps the conversation grounded in the facts.

At the close of the discussion, invite teachers to identify two values that they think are most centrally core, and two that they would want to shift their placement on the continuum. Revisit these in future department meetings or decision-making forums.

Perils, Pitfalls, and Other Helpful Notes

Discussing values can be challenging. It is not likely that all teachers will agree on the placement of a card on the continuum. This is an expected outcome—the process of considering and debating where the card should go, in our experience, surfaces critical issues and raises awareness of the extent to which a department does have shared values and the variance among teachers in how shared those values might be.

These conversations also have the potential to devolve into discussions about what students can and cannot do and about particular student populations (well, *your* students in Honors Algebra 2 might be able to do that, but *mine* in Track 4 Algebra 1 cannot). If the conversations go in this direction, it's a good time to remind all teachers that we are attempting to identify the extent to which values are universal—if they are true in some cases and not others, the center of the continuum is likely the right place for the card. Also try to redirect conversations from discussions of particular student groups, as the students are not there to have their own voices heard in such a conversation. As a follow up, one might invite teachers to make use of this activity with their students depending on the statements that are chosen.

Example statements for the continuum (some adapted from NCTM, 2014):

1. Equity—ensuring that all students have access to high-quality curriculum, instruction, and the supports they need to be successful—applies to all settings.
2. Equity is the same as equality.
3. Students possess different innate levels of ability in mathematics. Certain individuals are born with it while others aren't.
4. All students are capable of participating and achieving in mathematics, and all deserve support to achieve at the highest levels.
5. All students need to receive the same learning opportunities so that they can achieve the same academic outcomes.
6. Tracking promotes student achievement by allowing students to be placed in homogeneous groups where they can make the greatest learning gains.
7. Calculators and other tools should be used only after students have learned procedures with paper and pencil.
8. A deep understanding of mathematics is sufficient for effective teaching of mathematics.

9. The availability of open-source materials means that every teacher should design his or her own curriculum.
10. Effective mathematics instruction uses students' culture, conditions, and language to support and enhance learning.
11. Review and practice tests improve students' performance on high-stakes tests.
12. Effective teachers have a natural ability to provide innovative instruction that results in high levels of student achievement.
13. Effective teachers can work autonomously and in isolation. As long as the students are successful, all is well.
14. Students who are not fluent in English should be in a separate math class for English language learners.
15. The textbook should define the content and sequence for a class.
16. Math ability is a function of opportunity, experience, and effort, not innate intelligence.
17. The practice of low-level or slower-paced mathematics grouping should be eliminated.
18. Students who are learning English can learn the language of mathematics at grade level (or beyond) at the same time they are learning English.
19. Effective teachers become master teachers over time by continually improving their mathematical knowledge and skills.
20. All students are capable of making sense of and solving challenging mathematics problems and should be expected to do so.
21. Equity is attained when students receive the differentiated supports (time, instruction, curricular materials, programs) necessary to ensure that all students are successful.
22. Equity is only an issue for schools with racial and ethnic diversity or with significant numbers of low-income students.
23. Mathematics is independent of students' cultures, conditions, and language, and teachers do not need to consider these factors to be effective.
24. Some students just lack the cognitive, emotional, and behavioral characteristics to participate and achieve in mathematics.
25. Effective teaching practices have the potential to open up greater opportunities for higher-order thinking and for raising the mathematics achievement of all students.
26. Only high-achieving or gifted students can reason about, make sense of, and persevere in solving challenging mathematical problems.
27. Many more students need to be given the support, confidence, and opportunities to reach much higher levels of mathematical success and interest.

Activity 1.2: The No-Judgment Coffee-Hour Test Drive

The Goal

Engage in doing and thinking about mathematics together in ways that inform lesson planning and provide insights into student thinking.

The Task

Work together on a mathematical task that you will be giving students in the near future—preferably one that has been newly created or adapted from a resource. With your faculty, discuss the range of approaches that students might take—the strategies they might use, the representations they might produce (more and less useful), the tools you might make available, and the mathematical insights that are likely to come about from their work. If time allows, this can extend meaningfully into a discussion about the enactment of the task. The work of anticipating student thinking is the first of Smith and Stein's (2011) *Five Practices for Orchestrating Productive Mathematics Discussions,* and doing this work well informs the ways in which a teacher monitors student thinking as students work, selects and sequences student solution strategies, and makes meaningful connections to the underlying mathematical ideas.

Engaging in a task together can also lead to important conversations about how student thinking and learning is assessed. After working on the mathematics together, you might consider what you would see (in individual student work or a whole-class presentation) or hear (in small group work, in a whole-class presentation, in the questions students might ask) that would suggest to you that students were making sense of the important mathematical ideas.

Perils, Pitfalls, and Other Helpful Notes

The No-Judgment part of the activity title is important here. It is important that teachers are able to engage with the task—which may (perhaps should) represent mathematics that they have not taught in a while—with the notion that it is okay to have incomplete thinking, to struggle, and even to be wrong. This sort of thinking in fact, is an asset to these discussions, as it can help teachers to tune into the sort of thinking that they might encounter in their students.

If you do not have a good task at the ready for this activity, visit resources like Illustrative Mathematics (http://www.illustrativemathematics.org), the NCTM Illuminations website (http://illuminations.nctm.org), or the NCTM practitioner journals (Mathematics Teacher, Mathematics Teaching in the

Middle School). You can find many tasks there that are likely to connect to the mathematics that is coming up in a course or curriculum.

It is also helpful to find a relaxed venue for this work—depending on the mood of your school, the faculty room or departmental resource center might have too much baggage to afford the relaxed judgment-free discussion you would like to have. You might seek out a coffee shop or similar setting for the work. And making food available never hurts!

Activity 1.3: Math Teacher Circles

The Goal

Create a culture of regularly working on challenging mathematics problems with your professional colleagues.

The Task

Mathematics Teacher Circles (MTC) are a project of the American Institute of Mathematics, with support from the National Science Foundation, American Mathematical Society, Educational Advancement Foundation, the Mathematical Association of America, Math for America, and the National Security Agency. Mathematics Teacher Circles spun off from Mathematics Circles, which had the similar aim of gathering mathematicians and K–12 mathematics students together to collaborate on mathematics problems. The MTC aims to provide mathematics teachers with opportunities to experience mathematics as an open and creative endeavor - a stance that we are not often afforded as a teacher, where we are tasked with thinking through the math problems thoroughly before we engage students in them.

An MTC should run approximately 2 hours, and focus on a single challenging and open mathematics problem whose solution is not immediately obvious to the mathematics teachers and requires significant cognitive effort. The pedagogical stance of the circle should reflect a spirit of open inquiry, in which the facilitator(s) provide space for exploration, have good questions ready to ask to push thinking in meaningful ways, focus on mathematical practice, and not necessarily come to closure, such that participants walk away continuing to think about and puzzle over the problem. Teachers should be asked to consider a situation that can be mathematized in interesting ways, pose their own questions about the situation, and be encouraged to bring mathematical tools to bear in investigating those questions.

Finding a mathematician (or mathematicians) to partner with on regular Mathematics Teacher Circles is the optimal situation. A mathematician with a focus on K–12 education or preservice teacher preparation is more likely to be fluent in facilitating mathematics discourse and supporting genuine inquiry. The role of the mathematician is also to shape and focus the mathematical discussion with an eye towards moving teachers towards meaningful mathematics, and it is likely that many of these mathematical ideas will connect to the collegiate mathematics realm (frequently interesting and accessible topics like combinatorics, discrete mathematics, and graph theory).

The website www.mathteacherscircle.org serves as a hub for the MTC network. The site provides resources for organizing a circle, networking with other circles, identifying good problems to use, and facilitating the circle.

Perils, Pitfalls, and Other Helpful Notes

If you do not have a partnership already with a university or college, finding a good contact who shares an inquiry-based pedagogical approach might be a challenge. (See Activity 3.3, Fostering a Long-Term University Partnership, for some outreach recommendations.) Making use of the MTC Network via the website to find an existing circle nearby might be a good place to start, with the idea of spinning off your own circle later on if there is sufficient interest. You might also start by contacting a mathematics educator in the university (who may be housed in a School or College of Education) and work with them to identify a good mathematician to work with your circle. Service and outreach to the community is an important part of a faculty member's responsibilities at a university, so it is likely you'll be able to find a volunteer without having to dangle huge stacks of cash that most school districts have lying around for professional development. (Yes, that was a little bit of sarcasm, dear reader.)

Depending on the culture of your mathematics faculty, getting them to come willingly to the open inquiry might be challenging. The stance of the MTC is to promote an inclusive culture and to build community over the long haul. As such, a good starting approach would be to encourage the less resistant members of your faculty to engage in the first couple of MTC meetings and not worry about those who make the choice not to attend. As excitement about the MTC starts to increase with among faculty members that attend, it is likely that more reticent faculty members will want to see what the fuss is all about.

Activity 1.4: Three-Act Assessment Design

The Goal

Engage your colleagues in a richer, more interactive discussion of how students could be assessed.

The Task

We borrow this name from the much-ballyhooed Three-Act Tasks as originated by Dan Meyer. Meyer's Three-Act Tasks are based on a premise of a prelude that gets students interested in engaging in the mathematics and starts them posing important questions, transitions them into making use of meaningful mathematical tools to pursue the inquiry, and then both brings the mathematical ideas together in the third act while posing additional avenues for further discussion and exploration. We propose a similar Three Act structure for engaging your colleagues in discussion about assessment practices. This process has similarities to the ways in which the College Board goes about AP scoring, and other testing organizations like Educational Testing Service (ETS) develop exams like the Praxis series.

Act 1: Negotiating Content to Assess and What Performance Looks Like

This is a good entry point to the task, and frequently ends up being the start and the finish of a discussion about assessment among faculty teaching multiple sections of the same course. The spin on this task that may be unique is the second clause—determining what performance looks like. This goes beyond simply writing items and determining correct answers. The critical question to pose should be, what are the range of performances on this task/with this piece of content that would give us meaningful information about student thinking? A strong item should afford a range of performances, and those performances would provide important information. For example, a contextual problem might present an underlying quadratic situation and ask students to represent that situation at least two different ways and discuss the connections between them. A range of performances are possible on this task. Students may choose different representations, and their narrative about the connections between them may be more or less right, and may indicate different directionalities (e.g., starting with the symbolic and moving to a tabular representation; starting with a graph and using the graph to determine an equation). There may be multiple types of performances that would be considered as proficient for the mathematics content at play. This is perfectly acceptable and in fact, quite empowering for students in the end.

You will want to make sure that you record these ideas about what performance looks like in some way—chart paper, a whiteboard in a department meeting room, or (perhaps best) a collaborative electronic document. You will revisit this document in Act 3 and make revisions where appropriate.

Act 2: Taking the Assessment

Ask faculty to take the assessment and to work to produce a range of performances. This may be an activity to do together, or one to do outside of a regular meeting on one's own. Involving faculty who may not be teaching the content in question is valuable—this cross-training helps to build mathematics coherence across courses and raises the likelihood that unanticipated performances might arise if these teachers were not a part of Act 1. One might even recruit a few students to take the assessment or parts of it voluntarily (your Math Club, if you have one, would be a great choice). This may also be an opportunity to bring in a building principal and special education colleagues to take the assessment for a couple of reasons. First, they would provide another nonmathematics teacher data set to evaluate performance. Second, their pedagogical stances make it likely that they will provide some good critical assessment of the writing and structure of items. For example, a special education colleague might be more likely to identify a difficult piece of wording that whose meaning was implicitly obvious to the mathematics faculty.

The mathematics faculty taking the assessment who participated in Act 1 should have the idea of what performance looks like in mind and be open to the idea of iterating a second or third version of the "what performance looks like" document after taking the assessment. We recommend making any revisions that seem to be most critical based on the data before proceeding to Act 3. (Depending on the time and energy your faculty has, you could even engage in a fuller revision of the performance document at this time.)

Act 3: Administer the Assessment and Collaboratively Analyze the Data

Administer the assessment to students, and find a time to gather and look through the results. Break up student papers across faculty members and allow them to score the work of students that are not their own. Provide brief narrative feedback on items that lend themselves to such a response.

Gather the data across sections and teachers, and look at it in the aggregate. What do the data tell us about student performance? Where are we collectively strong, and where might we collectively have weaknesses in the content we are teaching? What aspects of student performance were we able to capture well? Were there aspects of student responses that we wanted to value but our rubric did not provide us an avenue? Pull examples of student work that provide responses for those questions, and engage in a discussion about how you might modify the performance standards to better capture those aspects of student thinking.

Perils, Pitfalls, and Other Helpful Notes

Time, time, and time. A full Three-Act assessment process takes energy before, during, and after a unit of instruction. It may end up causing a delay in getting assessments back to students, as an individual teacher may be able to complete the grading process more quickly on his or her own. It is possible, and perhaps even advisable, to separate the student assessment process from the process of revising the expectations. For example, teachers could use the criteria to assess student work and return the work to students. They could make copies of those scored assessments and convene with the department at a later time with those copies to analyze the assessment criteria and make changes. Keeping copies of student assessments, particularly digitally in the current technological age, is also just a great idea. Most current large-volume printers have a scanning option and make it easy to convert big stacks of student work into digital files. Make sure to remove student names or change them to numeric codes you can track to respect student privacy rights.

Another pitfall that may be less tangible is that this work can become very personal very quickly. Teachers are worried about the performance of their students, and looking at a broad set of student work across sections can bring to the surface repressed feelings of inequity with respect to students in sections, distribution and support of students with special needs or particular records of success, and even course loads and number of preparations per teacher. It can also result in teachers taking up a defensive posture, noting that they did not have time to teach one topic or another, had differential access to good materials for teaching, or were negatively impacted by conditions beyond their control (e.g., we had an assembly fifth period this week). In addition, teachers nearly always have their favorite aspects of content; personal pet peeves with respect to mathematics and how mathematics is done, written, and represented; and specific types and formats of questions. For example, they may value one type of solution strategy over another in terms of sophistication, have different views on the value of symbolic fluency, or even have some implicit bias regarding the neatness of handwriting and the organization of a written solution.

It is critical first to note that all of these tendencies are normal human nature and stem from a teacher's drive to do their very best for the students that they teach. It is easy to take such a discussion personally, as a good teacher feels a measure of responsibility for their students' learning and performance. This can be particularly prevalent in early-career teachers, who tend to be hyperfocused on their role in the classroom and tend to second-guess the decisions they make as having profound impacts on what students did or did not learn. (This is not to say that teacher decisions do not affect students' opportunities to learn, but early-career teachers tend to draw a straighter and bolder line between the two than is often warranted.) It is important in these cases to remind everyone of the goal—to develop better ways to assess student thinking and learning. There will always be variabilities in how and what one teacher teaches their students over another, and there will always be differences in the students in one section as compared to the other. The goal of this work is to provide as rich a set of opportunities for students to demonstrate mathematical performance in spite of those variabilities rather than excusing performance because of it.

At the start, one might not be able to do this for every unit all year long with faculty. That's okay! The Three-Act process can be used iteratively and over time. Choose a few units up front to use this process with, and the following year choose a different set. Over time, you will develop a rich, robust, and meaningful set of assessments that cover the entire high school mathematics spectrum.

Activity 1.5: Collaborative Observation

The Goal

Learn about the teaching practices of colleagues, provide evidence-based data and feedback to colleagues about teaching practice, open avenues for meaningful discussion of pedagogy.

The Task

Of the activities in this section, this is both the highest risk and highest reward. All too often, our teaching happens behind a closed door, with few opportunities to observe one another's practice. Even when we do have opportunities for peer observation, giving feedback can be particularly challenging. We often default to general evaluative feedback that is fairly mild (I like this, I might have changed this…) rather than using particular lenses to look at teacher-student interactions and engaging in an evidence-based discussion of what we notice and what we might wonder about.

Craig developed the a Collaborative Observation protocol (see Appendix A) making use of seven different lenses on teaching and learning—questioning, wait time, nature of content/academic culture, use of class time, voicing, task orientation, and content authority. The work was based on ideas gleaned from the teaching of the first field-based mathematics methods course at Michigan State (with influence from Artzt, Armour-Thomas, Curcio, & Gurl, 2015), and was shared with the Holt faculty and used within the district as a tool for analyzing and discussing teaching. We share it here as a tool for your use to bring focus and coherence to discussions about mathematics teaching and learning. For each of the seven lenses, a protocol describes an objective, evidence-based data collection method, such as observing student engagement at 5 minute intervals or noting times when the teacher takes (or misses) opportunities to explicitly point out the value of the mathematics at play in the lesson. These protocols and the associated data collection tables are designed to support colleagues in simply recording the events that transpired rather than interpreting or evaluating those events beyond the lens being used.

Ideally, an observation event would include several observers watching the same lesson and taking time together after the lesson to discuss what they have seen. (See the Perils, Pitfalls, and Other Helpful Notes for some perspectives on individual versus group observations.) This allows for multiple lenses and sources of evidence to be used in observing and reflecting on a lesson, which provides a more well-rounded view of the lesson and more material for a discussion of the lesson with the teacher.

After the observation by a group or an individual, the observer(s) and the instructor will want to meet to discuss the observation. The language of noticing and wondering (as described by Smith & Stein, 2011 among others) is a particularly helpful framing for this conference. Starting sentences with "I noticed..." and "I wondered..." keeps the discussion grounded in the actual events of the classroom and conjecturing about them rather than turning towards evaluation and judgment. In our experiences with both preservice and practicing teachers, these tools have been effective in keeping the conversation grounded in events and fostering productive interactions between instructors and observers.

Over the course of a school year, create a schedule in which the faculty can rotate through one another's classrooms, moving from instructor to observer and between different observer lenses. This should be viewed as a learning process for all, not just for the instructor—learning to observe using different lenses requires skill and practice, and there are some that may gravitate to one lens more than another. Building a flexible, fluent, and reflective observer base within the faculty is a critical component to moving instruction forward across your school.

The observation protocol we suggest using is included as Appendix A.

Perils, Pitfalls, and Other Helpful Notes

As noted above, the biggest challenge with collaborative observation is building a culture of critical colleagueship in which these observations can occur productively and without judgment. Depending on the culture of your building, it may help to bring in an outside person or persons to help get this work started. This might mean a mathematics coach from the district or a mathematics education colleague from a university. In the absence of such an outside influence, identifying colleagues to moderate the post-lesson observation debriefing between observers and the instructor is important. This person's sole role in the discussion will be to ensure that feedback remains objective rather than evaluative, and they can gracefully redirect or reorient conversations that drift towards evaluation. This in itself must not be judgmental—the slide towards evaluation is natural and in some cases, nearly inevitable—but simply a reminder of norms that have been agreed to with respect to the process.

The tension between individual and collective observation is important. Having any observer in the classroom can be anxiety-inducing for a teacher, and moving from a single observer to multiple observers can increase that anxiety. At the same time, a collective observation provides a richer picture of the lesson, and may help to avoid the notion that a single observer is unfair, biased, or in some other way out to get the instructor in question. One way to get started would be to hand-select members of the faculty who would be more amenable to initial observations and begin the process with those faculty members being observed (and we're guessing that you, dear reader, are on that short list). Another way to begin in a lower-risk environment is to use a video of a teacher publicly available online and use of the observation lenses as a group with the teacher's video. One can even simulate the observation conference by having a faculty member "play" the teacher on the video, responding as they might respond. Resources like Inside Teaching (http://insideteaching.org), Annenberg Media (http://www.learner.org/resources/browse.html?discipline=5), Video Mosaic (http://videomosaic.org), and the NCTM Principles to Actions Toolkit (http://www.nctm.org/ptatoolkit/) are excellent sources for lesson video. Many of these sources also include student work and lesson planning artifacts that can be used to better understand the teacher's instructional approach and the learning outcomes.

Another challenge that may be difficult to deal with directly is finding both the time and administrative support to release teachers to observe one another. From an administrative perspective, there is a very real cost involved, through either providing substitute teachers to release faculty to observe a single class, or to otherwise juggle student and teacher schedules to allow for an observation to occur. When pitching such a plan to an administrator, it is helpful to be prepared with information about the data you will collect from participants and how you will determine whether or not the work was worthwhile (and as such, worth the investment). Offer to collect reflections from teachers and perhaps interview the observed teacher a few weeks later to gather information about how their thinking and teaching may have changed based on the observation feedback. Invite administrators in on a session down the line (not the first one!) to see the process and the quality of feedback. Make connections to the process as helping to improve your local educator effectiveness/evaluation process ("this will likely lead to stronger teacher effectiveness outcomes"). But most of all, emphasize that professional learning is important because the end goal is to use these tools to improve student learning outcomes.

Finally, sustainability is an important factor to consider. Collaborative observation takes time and energy, and to sustain it over the long term requires faculty and administration to commit to making that time and preserving it. Having teachers opt out because they need to grade papers or advise a club will slowly erode away the culture of critical colleagueship. Do your best to ensure that this time is protected for all teachers, and that competing interests and activities can be set aside so teachers can focus on the work. This is an area in which a regular student teaching partnership with a local university can help. Observing and providing feedback is a critical part of the student teaching process, even more so today in an age where student teachers are required to submit high-stakes assessments like the edTPA assessment for licensure. To be strong mentors to student teachers, faculty must develop sound observational skills and practice giving feedback to those student teachers. As much as universities try to help their mentor teachers learn these skills, mentor teacher training by universities is frequently a one- or two-shot workshop before the start of a school year. Strengthening the faculty's ability to provide student teacher feedback will make your school a more productive environment for student teachers to learn and grow, and makes it more likely that promising young teachers will apply to join your faculty when jobs open in your building.

PART 2: COLLECTING AND ANALYZING CLASSROOM DATA

Activity 2.1: Demographics Analysis 1: Assessment Outcomes

The Goal
Identify patterns, trends, and biases in mathematics assessment outcomes.

The Task
Take as much of your state and local assessment data as you can, and analyze it demographically by gender, race, teacher, and year. The data from Holt provided in this volume, in addition to the data in *Embracing Reason* (Chazan, Callis, & Lehman, 2008) provide some examples of the sorts of data you might collect and the categories you might use to analyze these data.

Before you begin, identify the question at hand that you and your colleagues may be curious about. Consider this a research inquiry—as an educational researcher, Mike notes that his research is much more valid when he identifies his inquiry up front and then goes about determining how to collect and analyze data. When you start right with the analysis, you run the risk of reverse-engineering your question to fit the data. Frame your question as a good statistics hypothesis—it should not imply one outcome or another, but allow equally for either answer. Mike often uses the following example with his teachers conducting action research: "Why do calculators rot kids' brains?" is a biased question, even if it may be our gut feeling. "What influence does calculator use have on students' use of multiple mathematical representations?" is less biased and could be answered in a variety of ways. Once you have framed the question, consider the data that you could examine (whether it has already been collected or not) that could answer the question.

It is important to understand what these data sources can tell you and what they cannot. We're sure that many of you have experienced a data-driven presentation in your school or district where an administrator's conclusions about the data reach far beyond the data's scope and/or fail to account for the limitations of the data collection tools. (For example, I [Mike] once heard an administrator cite a single year's worth of state testing outcomes as evidence that a new curriculum was working—which had only been fully deployed to about 30% of the school's mathematics students. To boot, the assessment was given in October of that first year.) Prior to analyzing the data, write down the assumptions, conditions, and limitations about the assessments you will be using. If they are state assessments that favor more procedural outcomes, note that. If your school is only partially through an implementation of a new curriculum or new standards, write a brief narrative that provides that context. If you make use of course grades or common assessments, make notes about what aspects of that work teachers have the flexibility to change and adapt by classroom, and what aspects are truly common. This brief narrative introduction should not be difficult to write, but it is critical to have those caveats in mind up front as you proceed through the data analysis process. It helps to proactively take into account those factors rather than have to "make excuses for your data" after the analysis is complete.

Once you have created your narrative, identify the data sources and the variables you wish to disaggregate your data along. It is helpful at this point to think about how analysis results could be used and misused. For example, you may want to determine whether certain sections of classes are more successful than others. If this is a particular analysis worth exploring for your school, how could those data be misused? Could specific teachers end up being identified? Is there care that you could take to account for previous performance entering the year, particularly if classes are tracked at your school? Examining these questions will help all faculty members feel comfortable and safe with the results.

The report included in Appendix B (ACT vs. "true" assessment) is an example of such an inquiry that Craig produced for Holt using data from 2013.

Perils, Pitfalls, and Other Helpful Notes

As your faculty conducts your analysis, consider how you wish the results to be used. Is this analysis internal for your own reflection and improvement as a department? Is it something to share with administrators, the school board, or the community? Will it be posted or made available publicly? If a wider audience is implicated, more narrative work can and should be done around the report to help outsiders better understand your intentions for conducting the analysis, the analytical techniques used, and the limitations of the interpretations one can make for the data.

Working collaboratively on the analysis is best if your faculty can make the time and space to do so. This makes the process much more transparent, and also allows for all voices to give their input in the analytical techniques to be used. One can go down an almost infinite rabbit hole with respect to analytical sophistication—Should we use covariates? Create complex correlational models that integrate multiple factors at once like race, socioeconomic status, and prior achievement? What about more sophisticated techniques like hierarchical linear modeling? Choose a scope for your initial analysis and endeavor to stay within it—while there are always more statistical techniques one could use, there is also always more data in future years to be analyzed.

Activity 2.2: Demographics Analysis 2: Course-Taking Patterns

The Goal

Examine the relationships between the course offerings that are made available to students and the students who are enrolling and being successful in those courses.

The Task

As much as the negative impacts of tracking and grouping students by perceived ability have been documented, it is still a reality in a majority of high schools nationwide. One important step in making a case for dismantling tracking over time is to understand how the tracking structures affect students in your district. To do this, you will need to collect some systematic data about the demographics of your students and the courses that they take. This task is best accomplished with data over time—several years' worth of student course-taking data will guard against 1-year anomalies.

Once again, decide on the variables you wish to consider. You might analyze course-taking patterns by race, gender, socioeconomic status (free-and-reduced lunch status is often a good proxy for this), special education status, feeder district if you have more than one, transfer status, state test quartile or decile, or middle school achievement/last course. This analysis works best if you can track a cohort of students through your high school progress. This helps to determine if your course offerings are systematically impacting one group over another.

As an example, Figures 7.1 and 7.2 provide a course map with track enrollment by gender and race from an analysis of a district that Mike has worked with in the past. This very basic disaggregation of students by race and gender in the general and accelerated track shows some clear trends. At the middle school level, where students are first tracked, there are significantly more White students in the accelerated track as compared to the next largest race population, Black. As we move from Grade 6 into the high school, the ratio of White students in the accelerated track as compared to the general track increases from about 1.43:1 to nearly 2:1. The ratio of Black students in the accelerated track as compared to the general track moves from about 0.42:1 to 0.16:1. For this district, these data were compelling —systematically, their course structures were pushing more Black students out of the advanced track as they moved through the grades, and pulling more White students into the advanced track. Analyses related to economic advantage/advantage and disability status showed similar results—poorer students and students with disabilities were disappearing from the accelerated track as the grades increased (and fewer were entering the accelerated track at its start). These data together compelled the mathematics faculty and administration in this district to begin the process of dismantling their tracking structures and adopting a curriculum that was more accessible and equitable to all students.

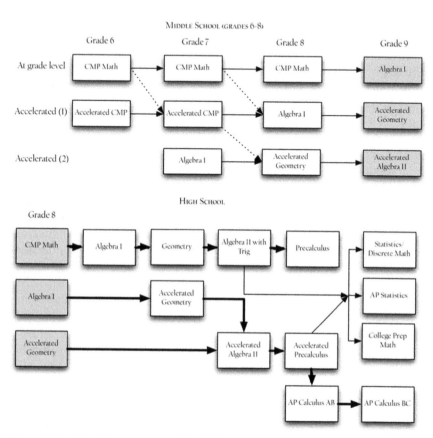

Figure 7.1. Sample course map describing tracking structures.

Track	Female	Male	Asian	Black	Hisp.	Am. Ind.	Pac. Isl.	Multi	White
Gr. 6 Gen.	116	126	13	60	15	0	1	13	140
Gr. 6 Accel.	119	141	8	25	9	0	0	18	200
Gr. 7 Gen.	120	116	8	72	14	2	0	30	110
Gr. 7 Accel.	134	132	7	14	7	0	0	20	218
Gr. 8 Gen.	118	139	8	72	14	2	0	30	110
Gr. 8 Accel.	121	112	7	14	7	0	0	20	218
Gr. 9 Gen.	149	159	9	95	13	2	1	21	167
Gr. 9 Accel.	128	106	4	10	9	0	0	23	188
Gr. 10 Gen.	154	179	20	82	20	2	1	21	187
Gr. 10 Accel.	96	83	6	11	5	0	0	14	143
Gr. 11 Gen.	154	179	9	104	13	5	0	23	237
Gr. 11 Accel.	96	83	1	5	3	0	0	4	102
Gr. 12 Gen.	154	179	6	50	16	3	0	13	76
Gr. 12 Accel.	96	83	10	8	3	3	0	10	141

Figure 7.2.   Tracking data disaggregated by gender and race.

One final idea to add in, borrowed from a district with which Mike had worked in years past, is to collect some qualitative data as well on students' experiences. This small district's mathematics curriculum director took time at the end of each year to interview every student for 15–30 minutes about their mathematics experience through the years in their district. This was an incredible data set that provided important data on how students experienced the mathematics curriculum, and represented an important effort to represent student voice into the district's data. Interviewing every student may not be feasible for a large district, but survey tools, representative interviews through stratified random sampling, and encouraging students to make video or audio reflections on their mathematics experience could be ways to manage a similar data collection effort in a larger district.

Perils, Pitfalls, and Other Helpful Notes

Most of these data are available already through your administrative office, so collecting these data and analyzing them should not be a terribly complicated matter. The biggest cautions regarding these data relate to after the fact—the outcomes of these analyses may be very difficult for faculty and administrators to face. In the data from the district above, I (Mike) was required to present these data to the mathematics teachers, central administration, and the school board. I had to think very carefully about how to talk about these data in a way that was honest about the systematic racial bias, did not point fingers at the mathematics faculty or the guidance department to suggest that this bias was intentional (I did not believe that it was), and provide thoughts about productive paths to take to solve the problem. This was a delicate dance, and there were some difficult feelings particularly amongst the faculty in facing these data head on. But ultimately, everyone saw the problem and once they worked through any initial bruised egos, realized they had to confront the problem clearly.

One final lesson learned from analyses like these is this—begin reporting internally. In many cases, the questions you're interested in answering, and the findings that the analysis will provide, will not necessarily be of broad interest to the community. The data cited earlier in this activity is a somewhat exceptional case, in that they clearly pointed to a change that was likely to be viewed as radical in the district. Limiting reporting to internal sources respects the professional work of teachers and administrators, even if public reporting will come later. In the case of the district described above, giving teachers and administration a chance to make sense of these results before they were presented in a board meeting open to the public allowed the faculty to think about the steps they wanted to take in response and present those steps in tandem with the public reporting of the data. This allowed them to shape the narrative immediately with the community and prevent a public backlash.

Activity 2.3: Developing Conceptual Understanding: Whole-School Student Assessment

The Goal

Sidestep the issue of a limited state assessment—choose a task that asks students to demonstrate conceptual understanding and administer it to all students at each grade level or course level.

The Task

You might have read the sentence in that goal and immediately recoiled. What a daunting task! Do we really have time to find a task for each grade level, administer it to all students, and make some sense of the outcomes? We're here to tell you that the answer to this is yes.

First, what you'll need to do is choose a good task that represents mathematics you'd expect your students to know at that grade level. The resources we've named previously like Illustrative Mathematics and NCTM Illuminations are great places to start. Make sure that you time the administration such that all students in the course/grade have had the opportunity to learn the content. For example, when we have done this in the past with middle and high school students, we have selected tasks for each of the middle grades, Algebra I, Geometry, and Algebra II. One could go further and create tasks for Precalculus and Calculus, but we have found that the Algebra II task tends to still yield interesting data for upper-division courses. The choice is yours.

Devise a simple set of rubrics for scoring the task. An example that Mike has used is shown in Table 7.1.[1] This basic rubric gave a score for the correctness of the mathematics in the task, and the extent to which the student communicated their mathematical understanding well. As can be seen from the rubric, your task does not have to assess every standard that students might learn, or even more than one. This rubric assesses one content standard and one Standard for Mathematical Practice from Common Core. The brief assessment provides a small snapshot into student performance at a particular moment in time.

Rubric for Delivery Trucks Task (Illustrative Mathematics, available at https://www.illustrativemathematics.org/content-standards/tasks/531)

Table 7.1.
Rubric for Delivery Trucks Task

	2	1	0
Content A-SSE.1.a Interpret expressions that represent a quantity in terms of its context.	Quantities for all 4 expressions correctly identified.	Some, but not all, quantities correctly identified.	No quantities correctly identified.
Communication SMP6 attend to Precision	All four quantities correctly and completely described, with correct units, distinguishing between average, total, and per trip amounts.	Some inaccuracies in units, or in description of quantities.	No attention to units or correct description of quantities.

We recommend administering the assessment to all students in a course at the same time. A single task like the Illustrative Mathematics example linked above would take students about 20 to 30 minutes to complete. In the grand scheme of the school year, taking 20 or 30 minutes to be able to collect some rich data from the entire student body is a worthwhile endeavor. This also can provide a very important formative assessment opportunity—just because the data are being aggregated across a larger data set does not preclude you as a teacher for using it towards your class's assessment dataset!

Scoring the data set can be as intensive or as relaxed as you make it. We have used this approach and scored every student paper in smaller settings. In larger settings, we have taken random samples of student work for each course or grade level and scored a set percentage. For a larger school, even taking 25% of the data set and scoring it might only take a few hours in the afternoon with the faculty pitching in.

You can look at the data within each course at the general level, or even find ways to disaggregate the data by teacher or one or more demographic dimensions. If you are not concerned as much about disaggregation, students do not even have to identify themselves on the paper. If you wish to engage in some disaggregation, assigning students numerical codes and having a key that links them to their demographics is a good approach. Student IDs that you may already have in your school could be used for this task.

Perils, Pitfalls, and Other Helpful Notes

As with any data set, understanding the limitations up front are important. For the example provided above, the data collected only gives a single snapshot of performance on one content standard and one standard for mathematical practice. Even still, if you have a state assessment that is not providing rich data (too procedural, does not break down by standard in a useful way), a small data collection effort like this can go a long way in understanding the sense students are making of their mathematical experience. For example, a district that Mike has worked with to administer a similar assessment with a district that had extremely strong standardized test scores. Their student work data, however, showed that most of their students scored a 1 or a 0 on their communication rubric related to the Standard for Mathematical Practice, even across courses and grade levels. Another district administered the Algebra II task to their Precalculus, Calculus AB, Calculus BC, and Calculus III students. Performance in the upper division classes was indistinguishable from the Algebra II students, with low scores on the Communication rubric across the higher level courses. These outcomes suggested a refocusing on communication and understanding was needed across the board.

Activity 2.4: Going for Broke: Commissioning an External Audit of Your Program

The Goal

Make use of district resources to support a broader data collection that encompasses aspects of all three of the previous activities (and more).

The Task

Activities 2.1, 2.2, and 2.3 are framed as tasks that a mathematics faculty can take on fairly manageably within a school or district. If your district can mobilize resources for a broader data collection that is not entirely a volunteer faculty effort, commissioning an external program audit is an excellent way to go. Have your district identify a team of university mathematicians, mathematics educators, and/or mathematics consultants (even administrators from a neighboring district) who understand your district and would be willing to engage in substantial data collection. Specify the goals of the mathematics audit, what data are and are not available (Will you ask them to make curricular recommendations or not? Will they have access to interview teachers and students? Observe classrooms?).

We recommend all three of the tasks in Part 2 be included as a baseline for an external audit. In addition, we suggest classroom observations, teacher interviews, student questionnaires interviews, a community survey, and a community open forum be considered for inclusion in an audit. One approach to observation and interviews is to strategically sample from the teachers in each building, grade level, and/or course. This may be a significant investment of money and time for the district, but an excellent district audit report from the outside can be a key catalyst for change, and takes some burden off the mathematics faculty to do large-scale data collection.

Perils, Pitfalls, and Other Helpful Notes

The biggest challenges here are fiscal and human resources. A good comprehensive audit can cost well into five figures for a good-sized district. Mike's most recent experience as a part of an audit team required about five hours a week for six months as one of a team of four conducting a full PK–12 audit. It was clear, however, that the 78-page report we produced was an important catalyst for ongoing change in the district and impacted decisions about curriculum, course offerings, professional development, and human resource deployment for years.

PART 3: ADVOCACY AND OUTREACH

Activity 3.1: Community Math Night 2.0: Focus on Process and Mindset

The Goal
Help the local community understand the sorts of mathematics your school and district is asking students to know and to do.

The Task
Community math nights have been a staple of ambitious districts, particularly in the elementary sphere, since the first wave of curricular reforms in the 1990s. One important goal of this work in elementary was to expose parents to the sort of sense-making tasks that we had been asking students to do as a ramp-up to fluency with the standard arithmetic algorithms.

At the high school level, asking parents and community members to come in and do content in high school mathematics may be daunting and may be a difficult sell for parents. A high school community math night instead might focus on processes and mindsets rather than specific mathematics content and procedures related to that content. Engaging the community in a single "low-threshold, high ceiling" task that can be approached multiple ways, with tools and manipulatives available, can highlight a number of key ideas. First, if it is facilitated effectively, parents who may not have productive mindsets towards mathematics can make progress on a mathematical task and feel as if they have the capability to know and do mathematics. Second, faculty facilitating the task can model research-based pedagogy through posing purposeful questions, facilitating meaningful mathematics discourse, encouraging participants to use and connect mathematical representations, and promoting productive struggle as participants work. Spending time after engagement with the mathematical task asking participants what supported their work and learning and identifying these teaching practices is necessary to bring to light the pedagogies that their students will be experiencing in their classrooms.

Perils, Pitfalls, and Other Helpful Notes
Good pedagogy will be the key to a successful community math night that is focused on process and mindset. It also may be helpful to review the research-based teaching practices from NCTM (2014) and the current research on mindset (Boaler, 2016; Dweck, 2006) at the close of the program to emphasize that this work represents research-based best practice.

As we described in Chapter 6, parents who have been through a traditional mathematics program and are in successful careers that use mathematics are often the toughest group to convince of the value of such an approach to teaching and learning mathematics. From their perspective, they were successful in a teacher-centered setting. It may be helpful to emphasize two things for this group. First, for every one parent who was successful, how many other parents are not using mathematics, were not successful in mathematics, or were/are scared of mathematics? You can encourage parents to think back to one of their mathematics classes and how many of their peers struggled. We can (and will) do better. Second, research does show that this approach generally does not hold back students at the top end of the achievement spectrum. The Holt data we have shared in this volume in fact, indicates that the number of students making it through advanced mathematics courses can increase, not decrease, when tracking is eliminated and a student-centered, meaning-making approach is adopted.

Activity 3.2: Business and Industry Idea Exchange

The Goal

Bring folks in from the corporate community to engage in discussions about how they use mathematics processes and mindsets in their work.

The Task

The idea of the guest speaker from industry has a long and, frankly, somewhat boring history in education. For decades we have sought to demonstrate the connections between our subjects and the real world by bringing professionals in to talk briefly about their work and its connections to mathematics. These connections are not always deep and substantive, as in our experience they have tended to focus on the content rather than processes,[2] practices, and mindsets. In addition, the mathematics that professionals often describe using is centered on arithmetic, which undercuts important messages about the utility of high school mathematics by portraying that useful math ends in sixth grade or so. Just like it is time for a change to community math nights, it's time for a change to the guest speaker trope.

Find a business or industry person that uses math in their work. Meet with them—preferably at their place of business—and spend some time experiencing how they actually make sense of mathematics and use mathematical processes in their work. Discuss with them what you notice, and how they can make connections to those ideas in a discussion with students. Explain to them the concept of growth mindset and ask them to help you understand how growth mindset is reflected in their daily work, human resource recruitment, training, or research and development. Coauthor an outline for engagement with your class.

In addition, help these folks identify an actual industry problem to bring to students—not to share the problem and solution, but to actually engage the students in working a bit on that problem and bringing their mathematical skills to bring to bear on it. By engaging students in a little bit of the actual work, they will have the opportunity to truly experience the ways in which mathematical processes and mindsets are used in their industry.

Perils, Pitfalls, and Other Helpful Notes

You are only limited here by the time and energy you can devote to finding a person, making visits, and identifying a good candidate. Leverage your own personal interests and those of your students here. For example, Mike has worked with the Baseball Research and Development staff of the Milwaukee Brewers baseball club to make such presentations to teachers and students. This connection came through reading about some of the advanced statistical measures the team was using to better quantify player performance, such as using piecewise trigonometric functions to find the range of any outfielder in any ballpark in the league. At first, the Brewers staff did not understand what they would have to say to teachers and students when Mike first approached them. Through a series of discussions, Mike shared what he was trying to help teachers and students see about statistics and problem solving, and the staff were able to connect their daily work to those ideas. Many industry folks are thrilled to engage with the K–12 world, and larger companies may even have incentives for such volunteer work to occur. All it takes is a little bit of reaching out and communication.

Activity 3.3: Fostering a Long-Term University Partnership

The Goal

Engage a nearby institution of higher education in collaboration with your mathematics faculty (ideally including placement of student teachers).

The Task

The partnership with Michigan State University was, and continues to be, a critical catalyst to Holt's Quiet Revolution. Not every school district is fortunate enough to have such a partnership evolve over time. A good first point of contact with a local university is the School or College of Education, and specifically the person at the school who arranges student teacher placements. This person knows all the faculty members with stakes in mathematics education, and frequently has the time and space in their day to help facilitate personal connections between your faculty and the university faculty. Using this entry point, find some time that fits faculty and staff schedules to go to the campus and meet to discuss your successes and challenges as a district. Listen to what faculty are working on in their research and service, and find common ground between faculty needs and district needs.

It is important here to start small. Even though one-shot professional development workshops do not lead to lasting change, they can be a part of a broader professional development strategy. Invite university faculty in to work with your faculty. Ask if there are colloquia or other learning experiences at the university that you could bring school and district faculty to attend. Ask about the possibility of a student teacher or field student being placed in your building, and connect the placement person directly with a teacher or two who would be a good fit. Speaking [Mike] as the supervisor of a secondary teacher certification program, I like to know a little bit about the teachers I'm placing my students with before I make the placement. This helps me understand both what the teacher's classroom practice might look like and how I can select a student whose strengths and challenges resonate well with that teacher.

After the small start, you can think about ways to move forward in broader ways. Getting to know the faculty in the Department of Mathematics can be more challenging—not all of them are always interested in the PK–12 world. Use your education colleagues to help you identify mathematicians that are interested in PK–12 and whose pedagogical skills resonate with a student-centered approach. Discuss broader partnership opportunities, such as professional development or research grants. Look for areas of intersection in research—if you are looking to improve outcomes with some demographic groups, and there is a researcher who focuses on teaching mathematics for social justice in their work, there is some natural common ground.

Perils, Pitfalls, and Other Helpful Notes

Fostering such a partnership is a long-term endeavor, and it often hangs on the stability of faculty at both institutions starting out. Funding is also a significant challenge. Higher education faculty have service to the professional community as a part of their charge—starting by appealing to their service interests is a good place to start. However, long-term sustained partnerships almost always include a research component in which faculty partner with the district to study teaching and learning. This can be daunting, particularly to a school administration that may not have experienced much research activity. Help administrators understand the importance of supporting faculty in their research in a way that also will provide the district with important information about teaching and learning.

Activity 3.4: Engagement in Assessment

The Goal

Engage mathematically-savvy members of the community in assessment judging.

The Task

As was described earlier in the book and also in *Embracing Reason* (Chazan, Callis, & Lehman, 2008), Holt adopted and refined the tradition of oral examinations as a part of student assessment over the years. This has been an important component at Holt for better understanding and honoring student competencies. A secondary outcome has been that it has allowed Holt to engage a wide range of mathematics stakeholders in the work of assessment. University faculty and staff, business and industry specialists, Board members, colleagues on their planning period, and other community members with mathematical expertise are regularly engaged as judges for the oral examination.

Whether you implement oral exams, project-based learning, or some other presentation-based form of assessment, inviting the public to both engage and to be a part of the work of assessing students is an excellent step. The number of questions I [Craig] have been able to respond to about the kids and our practice from judges has given me confidence to claim that there may be no better public relation. Having instructions, rubrics, and an on-boarding process is critical to make students and judges feel comfortable with the work.

There are many variations to how this may be implemented. At Holt, we have done it for final exams (so, in January and then again in June), but there is no reason to avoid unit assessments, midterms, or projects that run in addition to other tests. And even the structure for how students will present to judges can vary wildly. Mike Lehman used to set a group of students with two to three judges and each one in turn would explain a project, chosen at random by the judges, they had completed earlier in the term. Sometimes he would also do a group problem that the students would collaboratively solve in front of the judges. I co-opted a scheme from Kelly Hodges and Sandy Callis where a group of students rotates to a table of one to three judges who are responsible for one of a handful of questions students have prepared in advance. One predetermined student is assigned to go so that each student ends up presenting two or three of the six to eight exam problems to different sets of judges over the course of eight 10-minute rotations. The options are limited only by the timeframe and human resources.

Perils, Pitfalls, and Other Helpful Notes

Setting this up requires quite a bit of front-loaded work: inviting judges, keeping track of who can commit to what time periods (if doing over multiple sessions or days), generating good questions, giving the students time to prepare, writing and sharing the rubric, writing notes and possible solutions for judges ahead of time, creating the rotations and directions for students and judges, making copies, and so forth. You should probably also have a back-up plan in case a judge does not show or gets called away (often a building administrator might need to handle something last-minute). But unless the alternative is a scanned multiple-choice exam, then some of the time is recouped afterwards when your main task is to average judges' scores and copy the feedback for students.

You also have to be comfortable taking a leap of faith to allow your students, and by extension your teaching, be on full display. However, it does not have to be pretty, or polished; in fact, I tell judges that some students may not have made good choices to prepare or may try to try and "talk the talk" by parroting someone else's thinking and their job is to determine who understands and to what degree. In fact, people not often used to grading students tend to overinflate the grade and so having a rubric and a description for the judge about what different levels of sophistication looks like is particularly useful. And do not forget, you can control how much you let the grade from the judges figure in, and if you want/need other components.

Just as we discussed with collaborative observation, stakeholders engaged in assessment practices might gravitate towards general weak praise in an effort not to hurt students' feelings. Emphasize that they should provide substantive feedback and give them specific tools, including modeling the language they should use, to help them engage more substantively with students. You might also provide stakeholders with some context about what students have learned and what they can expect strong (and weaker) performance to look and sound like.

NOTES

1. Our thanks to Mike's colleague in the UWM Mathematics Department, Dr. Kevin McLeod, for his development of this rubric.
2. When we talk about mathematical processes and practices, there are a number of ways those processes and practices could be characterized. One way that has currency in most places at the time of writing are the Standards for Mathematical Practice. They are, make sense of problems and persevere in solving them, reason abstractly and quantitatively, construct viable arguments and critique the reasoning of others, model with mathematics, use appropriate tools strategically, attend to precision, look for and make use

of structure, and express regularity in repeated reasoning. There are other descriptions, such as the Mathematical Habits of Mind, and likely to be even more in the future. The specific process is less important than the idea that mathematics is about process rather than just content.

APPENDICES

APPENDIX A:
HHS MATH DEPARTMENT COLLABORATIVE
OBSERVATION SYSTEM

Created 2013, with influence from Artzt, Armour-Thomas, Curcio, an Gurl (2015) observation suggestions for preservice teachers, and Vacc's (1993) and Boaler & Brodie's (2004) classifications of questions.

The observers listed below are not prioritized—the options are dependent on the number of people observing and what the teacher is most interested in getting information about. Some roles may be combined and done by one person. The role of the observer is to take accurate data about their specific facet of the classroom experience to (a) inform his or her own practice and (b) share with the teacher for their own reflection. Ideally, there would be several observers examining the same lesson, and time to reflect together on what they saw; like a Lesson Study, but with a focus on pedagogy rather than the lesson itself.

A Quiet Revolution:
One District's Story of Radical Curricular Change in High School Mathematics, pp. 177–193
Copyright © 2018 by Information Age Publishing

Observer	Focus	Description
Observer #1	Questioning	Records the type of questions s/he hears the teacher ask, in what context, and notes whether it was choral response, voluntary, or teacher-selected/random.
Observer #2	Wait Time	For each question asked, this observer uses a stopwatch to record the actual length of wait time after a question is asked and again once a response is given.
Observer #3	Nature of Content/ Academic Culture	Observe for times when the teacher takes or misses opportunities to explicitly point out the value of math *and* times when the teacher reinforces or undermines the enhancement of an academic culture.
Observer #4	Use of Class Time	Observer keeps a running list of the classroom activities with the time. As this is being created, s/he is taking note of where the teacher is in the room, what the majority of students are doing, and what phase of the lesson is occurring.
Observer #5	Voicing	The observer uses a seating chart (or makes one) and records the obvious attributes of the students (gender, cultural background, classroom status, etc.) and then codes all of the occurrences and types of student interaction.
Observer #6	Task Orientation	At 5 minute intervals, the observer notes what is happening in class and counts the number of students on task vs. the number off task.
Observer #7	Content Authority	Observer records places where the teacher is the content authority (either by telling or asking leading questions) or when s/he avoids doing the thinking for the students (i.e., asks a question back).

After the observation, team breaks to compile information, make graphs, and so forth, and reconvenes to discuss what they noticed; ideally on the same day. When possible, a video placed on a tripod in the back of the classroom may also be used for review later.

Observer #1:   **Questioning**

Date: _____   Teacher: _____

Course: _____   Unit: _____

Lesson Topic: _____

Observer Name: _____

Directions: Listen carefully anytime you hear a question posed by the teacher. Based on the descriptions below, choose the code that best describes it, and keep a record of the number of each type. Keep the questions separated by whether they were asked in a whole-group setting, posed to a small group only, or to an individual student. For those in a whole-class setting, note if it was asked in such a way to elicit voluntary response, choral response, or teacher-selected/random responses.

After the observation, make a frequency bar graph for the types of questions asked in the whole class setting, the groupwork setting, and when talking to individuals. Also report out what percent of the whole-class discussion questions were requesting voluntary, choral, or teacher-selected responses.

Note: Sometimes, statements could be considered questions, and vice versa.

Question Type	Example(s)	Code
Low-level	Questions that have a factual, or one, specific answer: "What do we call this kind of graph again?" "Do we add or subtract here?"	L
Open, Observation	Open-ended, answer not pre-determined: "What do you notice about...?"	O
Open, Reasoning	Open-ended, asking for students' reasoning: "Why does...?"	R
Probing	Questions to get further at a student's response: "Say more...."	P
Establishes Context	Questions meant to clarify the situation: "Does anyone's family own a business?" "How many players does a softball team need?"	C
Generating Discussion	Questions meant to engage or elicit responses: "Does anyone have a different opinion?"	D

Put a tally mark in the appropriate place in the matrix below as you observe for teaching questioning. The last row is for questions that may not fit into the other categories. In the last column at least, instead of tallies, use *v* for voluntary response questions, *s* for selected response, and *c* for choral response.

	To individual	To a group	Whole-class
L			
O			
R			
P			
C			
D			
?			

Observer #2:   **Wait Time**

Date: _____   Teacher: _____

Course: _____   Unit: _____

Lesson Topic: _____

Observer Name: _____

   Directions: Listen carefully for any time the teacher asks a question in the whole-group setting, and with a stopwatch record below the amount of time (in seconds) allowed before a student is given permission to answer *and then* the amount of time after the answer is given before the teacher or another student is allowed to respond. After the lesson, compute the average wait-time for both before and after.

Question #	Wait-Time (sec)	Wait-time Following Response
1		
2		
3		
4		
5		
6		
7		
8		
9		
10		
11		
12		
13		
14		
15		
16		
17		
18		
19		
20		
21		
22		
23		
24		
25		
26...		

Observer #3:   **Nature of Content & Academic Culture**

Date: _____   Teacher: _____

Course: _____   Unit: _____

Lesson Topic: _____

Observer Name: _____

Directions: Take notes throughout the class period for (a) times when the teacher **takes** or **misses** opportunities to explicitly point out the value of mathematics, and (b) when the teacher **reinforces** or **undermines** the enhancement of an academic culture. Be sure to be attentive both <u>positive and negative</u> teacher moves that could have an impact on the norms of the classroom.

Observer #4:   **Use of Class Time**

Date: _____   Teacher: _____

Course: _____   Unit: _____

Lesson Topic: _____

Observer Name: _____

Directions: In the left margin, keep track of the times in which class-room activities occur. Begin with the time of the bell and end with the bell, creating a running log of what is happening at what times: what phase of the lesson is occurring, the location of the teacher in the room, what the majority of the students are doing.

Time	Lesson Phase	Teacher Location	Description of What the Majority of Students Are Doing

Observer #5:   **Voicing**

Date: _____   Teacher: _____

Course: _____   Unit: _____

Lesson Topic: _____

Observer Name: _____

Directions: To start, you will need a seating chart of the room, and quickly record the obvious attributes of the students that are present (gender, cultural background, classroom status, etc.). Using the codes below, record the ways in which students are using their voices during class:

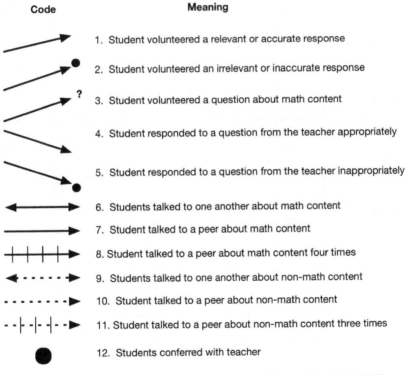

Code	Meaning
	1. Student volunteered a relevant or accurate response
	2. Student volunteered an irrelevant or inaccurate response
	3. Student volunteered a question about math content
	4. Student responded to a question from the teacher appropriately
	5. Student responded to a question from the teacher inappropriately
	6. Students talked to one another about math content
	7. Student talked to a peer about math content
	8. Student talked to a peer about math content four times
	9. Students talked to one another about non-math content
	10. Student talked to a peer about non-math content
	11. Student talked to a peer about non-math content three times
	12. Students conferred with teacher

Adapted from Artzt (2008)

Observer #6:   **Task Orientation**

Date: _____   Teacher: _____

Course: _____   Unit: _____

Lesson Topic: _____

Observer Name: _____

Directions: At 5 minute intervals, note what is happening in class and count the number of students on task vs. the number off task.

Increment	Clock Time	General Description of What's Occurring	# On Task	# Off Task
0:00-0:05				
:05-:10				
:10-:15				
:15-:20				
:20-:25				
:25-:30				
:30-:35				
:35-:40				
:40-:45				
:45-:50				
:50-:55				
:55-:60				

Afterwards, on graph paper, make a graph of the "percent on task" at 5-minute intervals. Label on the graph when important transitions occurred.

Observer #7:   **Content Authority**

Date: _____   Teacher: _____

Course: _____   Unit: _____

Lesson Topic: _____

Observer Name: _____

Directions: Observe carefully for times during the lesson when the teacher indicates that s/he is the authority for content knowledge. Examples may include:

- asking leading questions, where it is clear to students that s/he already knows the answer
- the teacher telling, lecturing, or explaining a concept for them
- affirming or correcting student responses
- subtle or obvious gestures/facial expressions/voice inflection that indicate correctness

Also, note places where the teacher skillfully avoids doing the thinking for the students. Examples may include:

- redirects students to each other or their notes
- asks a question back or invites them to say more about their thinking
- goes with a student error until they either correct it themselves or it reaches a conclusion that doesn't pan out
- acknowledging student conjectures (right or wrong)

APPENDIX B: DATA SUMMARY REPORT— TO WHAT EXTENT IS THE ACT A TRUE ASSESSMENT AS COMPARED TO COURSE GRADES?

Date completed: *12/3/2013*

Analysis and Write-up: *Craig Huhn, on behalf of the district data team*

Context

Due to ease and accessibility, the ACT is commonly used when trying to determine if enough students know enough stuff. However, some questions have been raised about the ability of this test to actually measure student understanding of required course content. Reasons cited include:

- Teachers examining released and/or practice items
- A history of stable scores even though the courses, expectations, students, times, and teachers have changed significantly
- A confoundance with other variables that impact success on the ACT (we know that ACT math scores correlate *highly* with the percent free-and-reduced numbers, for example)
- ACT is a normed multiple-choice timed test designed to separate and sort

The conjecture is that the ACT scores (in mathematics, at least) measure cleverness and reading comprehension, and not how many students understand how much of the content standards. The implications of this would be that the ACT would potentially be an inappropriate metric to use, if we are curious about what students know what required content to what degree.

Question

Since teachers notice many students who demonstrate amazing understanding but earn low ACT scores, and vice versa, the question is, how wide-spread is this phenomenon? In other words, *can ACT scores be used as a proxy for evaluating how much content a student really understands?* This query is specifically examining this with respect to mathematics.

Metrics

a.   What we really want to know is if the ACT score correlates with the score students were given after rigorous testing by their teacher. Therefore, the ACT score in math, paired with that student's math grades will be the data requested. Because grades can be subjective and in the control of the teacher with immense pressure to "move kids through" and keep failure rates low, it was decided that Final Exam grades would be the best measure, as opposed to the final term grades. In mathematics, these are departmental exams, with common grading for consistency across the board. Algebra C was chosen because it is the culmination of a three semester examination of functions in the Algebra sequence, and it is the final most recently taken by juniors taking the ACT.

b.   In order to determine a relationship, a threshold for the correlation coefficient of .8 for data of this nature (social and multivariate) would be considered "high." Correlation of .7 would be decent (implying that 49% of the variation in ACT scores are explained by the student's grade), and anything below .6 could be cause to say that there is no relationship between the variables beyond a simple "better students generally do well in school and on the ACT overall."

Assumptions

1.   Because the mathematics department has put a high value on the summative assessment as an opportunity for students to show what they have learned, they have worked to create a final exam that is aligned to the big ideas of the course, has been vetted for several years, and acquires evidence of deep understanding. Rather than put together a quick test to satisfy Board requirements or get it graded and out of the way as fast as possible, the department has worked to create an exam that gets at true understanding of the course content via large constructed response questions, short answer, and some multiple choice (at the time, a requirement to include). Additionally, the department collaboratively grades the constructed response, modeled after the AP grading process with consistent expectations/rubric. Therefore we are taking as an assumption that the final exam is a legitimate measure of student understanding.

2. We are also operating under the assumption that grades submitted via Skyward by teachers are the legitimate comparable scores on the final (that the common scores were not altered by the teacher or someone else with access later). This should also be the case.

Cautions

Whenever correlation is being looked at, it is important to keep in mind that correlation does not indicate causation.

Any results found here may or may not be true in other subject areas. Further study would be required.

Analysis

The population is all HHS students that took the ACT last year and who have a grade in Algebra C ($n = 245$).

The scatterplot of the data is given:

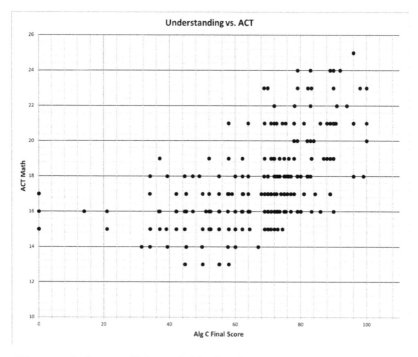

The correlation coefficient of this data is $r = 0.551144$.

R^2 is therefore 0.30376.

Conclusions

Both visually and quantitatively, it appears that the ACT is a poor predictor for content understanding. The correlation coefficient is far below the threshold set. The coefficient of determination indicates that only 30% of the variation in ACT score can be attributed to what students know, as evidenced on the course final.

REFERENCES

ACT. (2015). Preparing for the ACT Test. Retrieved on August 15, 2017, from http://cdn2.hubspot.net/hubfs/360031/ACT-2015-16.pdf?t=1470774914678

American Statistical Association. (2014, April 8). ASA Statement on using value-added models for educational assessment. Retrieved August 10, 2017, from http://www.amstat.org/asa/files/pdfs/POL-ASAVAM-Statement.pdf

Artzt, A. F., Armour-Thomas, E., Curcio, F. R., & Gurl, T. J. (2015). *Becoming a reflective mathematics teacher: A guide for observations and self-assessment*. New York, NY: Routledge.

Association of Mathematics Teacher Educators. (2017). *Standards for Preparing Teachers of Matheamtics*. Retrieved on September 29, 2017, from http://amte.net/standards

Baker, E. L., Barton, P. E., Darling-Hammond, L., Haertel, E., Ladd, H. F., Linn, R. L., ... & Shepard, L. A. (2010). Problems with the Use of Student Test Scores to Evaluate Teachers (EPI Briefing Paper #278.) *Economic Policy Institute*. Retrieved from http://www.epi.org/publication/bp278/

Ball, D. L., Thames, M. H., & Phelps, G. (2008). Content knowledge for teaching: What makes it special? *Journal of Teacher Education, 59*(5), 389–407.

Berkshire, J. (2017, August 7). The right wing in America has long tried to destroy "government schools." *Salon Magazine*. Retrieved on August 10, 2017, from http://www.salon.com/2017/08/07/the-right-wing-in-america-has-long-tried-to-destroy-government-schools

Black, P., & Wiliam, D. (2010). Inside the black box: Raising standards through classroom assessment. *Phi Delta Kappan, 92*(1), 81–90.

Boaler, J. (2016). *Mathematical mindsets: Unleashing students' potential through creative math, inspiring messages, and innovative teaching*. San Francisco, CA: Jossey-Bass.

Boaler, J., & Brodie, K. (2004, October). The importance, nature and impact of teacher questions. In D. E. McDougall & J. A. Ross (Eds.) *Proceedings of the twenty-sixth annual meeting of the North American Chapter of the International Group for the Psychology of Mathematics Education* (Vol. 2, pp. 774–782). Toronto, Canada: Ontario Institute for Studies in Education.

Boston, M. D., & Smith, M. S. (2009). Transforming secondary mathematics teaching: Increasing the cognitive demands of instructional tasks used in teachers' classrooms. *Journal for Research in Mathematics Education, 40*(2), 119–156.

Chazan, D. (2000). *Beyond formulas in mathematics and teaching: Dynamics of the high school algebra classroom.* New York, NY: Teachers College Press.

Chazan, D., Callis, S., & Lehman, M. (2008). *Embracing reason: Egalitarian ideals and the teaching of high school mathematics.* New York, NY: Routledge.

Cohen, E. G., & Lotan, R. A. (2014). *Designing groupwork: Strategies for the heterogeneous classroom* (3rd Ed.). New York, NY: Teachers College Press.

Conference Board of the Mathematical Sciences. (2001). *The mathematical education of teachers* (Vol. 11). Washington, DC: American Mathematical Society.

Conference Board of the Mathematical Sciences. (2012). *The mathematical education of teachers II.* Washington, DC: American Mathematical Society.

Consortium for Mathematics and Its Applications (COMAP) & Society for Industrial and Applied Mathematics (SIAM). (2016). *Guidelines for assessment & instruction in mathematical modeling education.* Bedford, MA & Philadelphia: Authors.

DuFour, R. (2002). The learning-centered principal. *Educational Leadership, 59*(8), 12–15.

Dweck, C. S. (2006). *Mindset: The new psychology of success.* New York, NY: Ballantine Books.

Fay, J., & Funk, D. (1995). *Teaching with love and logic: Taking control of the classroom.* Golden, CO: Love and Logic Press.

Featherstone, H., Crespo, S., Jilk, L. M., Oslund, J. A., Parks, A. N., & Wood, M. B. (2011). *Smarter together! Collaboration and equity in the elementary math classroom.* Reston, VA: National Council of Teachers of Mathematics.

Ferro, S. (2013, September 12.) Everything you've ever been told about how you learn is a lie. Retrieved from http://www.popsci.com/science/article/2013-08/everything-youve-ever-been-told-about-how-brain-learns-lie

Greeno, J., & the Middle School Mathematics through Applications Project. (1998). The situativity of knowing, learning and research. *American Psychologist, 53*(1), 5–26.

Grouws, D. A., Tarr, J. E., Chávez, Ó., Sears, R., Soria, V. M., & Taylan, R. D. (2013). Curriculum and implementation effects on high school students' mathematics learning from curricula representing subject-specific and integrated content organizations. *Journal for Research in Mathematics Education, 44*(2), 416–463.

Hodgen, J., & Wiliam, D. (2006). *Mathematics inside the black box: Assessment for learning in the mathematics classroom.* Brentford, UK: Granada Learning.

Huhn, C. (2005). How Many Points Is This Worth?. *Educational Leadership, 63*(3), 81–82.

Huhn, C., Huhn, K., & Lamb, P. (2006). Lessons teachers can learn about students' mathematical understanding through conversations with them about their thinking: Implications for practice. In Van Zoest, L. (Ed.), *Teachers engaged in research: Inquiry into mathematics classrooms, Grades 9–12*. Charlotte, NC: Information Age Publishing.

Humphreys, C., & Parker, R. (2015). *Making number talks matter: Developing mathematical practices and deepening understanding, grades 4–10*. Portland, ME: Stenhouse.

Ingham Clinton Education Association. (2008). Master agreement between the Ingham Clinton Education Association and the Board of Education, July 1, 2008-June 30, 2010. Unpublished document.

Jamison, N. (2015, October 28). These are the 10 happiest cities in Michigan. Retrieved from https://www.homesnacks.net/these-are-the-10-happiest-cities-in-michigan-123155/.

Koretz, D., Mitchell, K., Barron, S., & Keith, S. (1996). *The perceived effects of the Maryland school performance assessment program*. Los Angeles, CA: Center for Research on Evaluation, Standards, and Student Assessment (University of California at Los Angeles).

Kimball, M., & Smith, N. (2013, October 28). The myth of "I'm Bad at Math." *The Atlantic*. Retrieved from https://www.theatlantic.com/education/archive/2013/10/the-myth-of-im-bad-at-math/280914/?utm_source=atlfb

Kroll, Andy. (2014, January/February). Meet the new Kochs: The DeVos clan's plan to defund the left. *Mother Jones*. Retrieved August 10, 2017 from http://www.motherjones.com/politics/2014/01/devos-michigan-labor-politics-gop/

Lane, S. (2004). Validity of high-stakes assessment: Are students engaged in complex thinking? *Educational Measurement: Issues and Practice*, *23*(3), 6–14.

Lewis, A. C. (2002). Washington commentary: School reform and professional development. *Phi Delta Kappan*, *83*(7), 488.

Lotan, R. A. (2003). Group-worthy tasks. *Educational Leadership*, *60*(6), 72–75.

McCaffrey, D. F., Hamilton, L. S., Stecher, B. M., Klein, S. P., Bugliari, D., & Robyn, A. (2001). Interactions among instructional practices, curriculum, and student achievement: The case of standards-based high school mathematics. *Journal for Research in Mathematics Education*, 493–517.

McKnight, C. C., Crosswhite, F. J., Dossey, J. A., Kifer, E., Swafford, J. O., Travers, K. J., & Cooney, T. J. (1987). *The underachieving curriculum: Assessing US school mathematics from an international perspective* (A National Report on the Second International Mathematics Study). Champaign, IL: Stipes.

McLeod, D. B., Stake, R. E., Schappelle, B. P., Mellissinos, M., & Gierl, M. J. (1996). Setting the Standards. In S. A. Raizen and E. D. Britton (Eds.), *Bold ventures: Case studies of U.S. innovations in mathematics education* (Vol. 3, pp. 13–132). Norwell, MA: Kluwer Academic Publishers.

Michigan Enrolled House Bill No. 4627. (2011). Michigan Public Act 102. Retrieved on August 15, 2017 from https://www.legislature.mi.gov/documents/2011-2012/publicact/htm/2011-PA-0102.htm

Michigan Enrolled House Bill No. 4929. (2012). Michigan Public Act 53. Retrieved on August 15, 2017, from https://www.legislature.mi.gov/documents/2011-2012/publicact/pdf/2012-PA-0053.pdf

Michigan Public School Employees Retirement System Reform. (2012). Michigan Public Act 300. Retrieved on August 15, 2017, from http://www.michigan. gov/orsschools

Michigan Right-to-Work Reform. (2012). Michigan Public Act 349. Retrieved on August 15, 2017, from http://www.michigan.gov/orsschools

Michigan Publicly Funded Health Insurance Contribution Act. (2011). Michigan Public Act 152. Retrieved on August 15, 2017 from http://legislature.mi.gov/ doc.aspx?mcl-act-152-of-2011

Michigan Senate Bill 0103. (2011) Michigan Public Act 175. Retrieved on August 15, 2017 from http://www.legislature.mi.gov/documents/2015-2016/publi-cact/htm/2015-PA-0173.htm

mischooldata.org (2017, February 27). District/School Information. Retrieved on February 27, 2017, from mischooldata.org

Moses, R. P., & Cobb, C. E. (2001). *Radical equations: Math literacy and civil rights.* Boston, MA: Beacon Press.

National Center for Education Statistics (NCES). (2009). *NAEP 2008: Trends in Academic Progress. NCES 2009–479.* Washington, DC: Author.

National Commission on Excellence in Education. (1983). A nation at risk: The imperative for educational reform. *The Elementary School Journal, 84*(2), 113–130.

National Council of Supervisors of Mathematics. (2008). *The PRIME Leadership Framework.* Bloomington, IN: Solution Tree.

National Council of Supervisors of Mathematics. (2014). *It's TIME: Themes and Imperatives for Mathematics Education.* Bloomington, IN: Solution Tree.

National Council of Teachers of Mathematics (1980). *An agenda for action: Recommendations for school mathematics of the 1980s.* Reston, VA: Author.

National Council of Teachers of Mathematics (1989). *Curriculum and evaluation standards for school mathematics.* Reston, Virginia: Author.

National Council of Teachers of Mathematics. (1991). *Professional standards for teaching mathematics.* Reston, VA: Author.

National Council of Teachers of Mathematics. (1995). *Assessment standards for school mathematics.* Reston, VA: Author.

National Council of Teachers of Mathematics (2000). *Principles and standards for school mathematics.* Reston, VA: Author.

National Council of Teachers of Mathematics (2009). *Focus in high school mathematics: Reasoning and sense making.* Reston, VA: Author.

National Council of Teachers of Mathematics (2014). *Principles to actions: Ensuring mathematical success for all students.* Reston, VA: Author.

National Governors Association Center for Best Practices, & Council of Chief State School Officers. (2010). Common Core State Standards for mathematics. Retrieved from http://www.corestandards.org/Math

National Research Council. (2000). *How people learn: Brain, mind, experience, and school: Expanded edition.* Washington, DC: National Academies Press.

National Research Council & Mathematics Learning Study Committee. (2001). *Adding it up: Helping children learn mathematics.* Washington, DC: National Academies Press.

O'Connor, K., & Wormeli, R. (2011). Reporting student learning. *Educational Leadership, 69*(3), 40–44.

Olsen, C. (2014, April 25). Best places for home ownership in Michigan. Retrieved from https://www.nerdwallet.com/blog/mortgages/home-search/best-places-homeownership-michigan/

O'Neil, C. (2016). *Weapons of math destruction: How big data increases inequality and threatens democracy.* New York, NY: Crown.

Parke, C. S., Lane, S., & Stone, C. A. (2006). Impact of a state performance assessment program in reading and writing. *Educational Research and Evaluation, 12*(3), 239–269.

Post, T. R., Medhanie, A., Harwell, M., Norman, K. W., Dupuis, D. N., Muchlinski, T., ... & Monson, D. (2010). The impact of prior mathematics achievement on the relationship between high school mathematics curricula and postsecondary mathematics performance, course-taking, and persistence. *Journal for Research in Mathematics Education,* 274–308.

Rasmussen, S. (2015, March 11). Why the Smarter Balanced Common Core math test is fatally flawed. *EdSurge.* Retrieved on September 27, 2017 from https://www.edsurge.com/news/2015-03-11-why-the-smarter-balanced-common-core-math-test-is-fatally-flawed

Reys, B. (Ed.). (2006). *The intended mathematics curriculum as represented in state-level curriculum standards: Consensus or confusion?* Scottsdale, AZ: Information Age Publishing.

Schmidt, W. H., Houang, R., & Cogan, L. (2002). A coherent curriculum: The case of mathematics. *American Educator, 26*(2), 10–26, 47–48.

Schoenfeld, A. H. (2004). The math wars. *Educational policy, 18*(1), 253–286.

Senk, S. L., & Thompson, D. R. (Eds.). (2003). *Standards-based school mathematics curricula: What are they? What do students learn?* Mahwah, NJ: Lawrence Erlbaum.

Shulman, L. S. (1986). Those who understand: Knowledge growth in teaching. *Educational researcher, 15*(2), 4–14.

Smith, M. S., & Stein, M. K. (2011). *Five practices for orchestrating productive mathematics discussions.* Reston, VA: National Council of Teachers of Mathematics.

Spillane, J. P. (2000). Cognition and policy implementation: District policymakers and the reform of mathematics education. *Cognition and instruction, 18*(2), 141–179.

Stecher, B. M., Vernez, G., & Steinberg, P. (2010). Accountability for NCLB: A report card for the No Child Left Behind Act. Retrieved August 10, 2017 from https://www.rand.org/pubs/periodicals/rand-review/issues/summer2010/nclb.html

Stein, M. K., & Lane, S. (1996). Instructional tasks and the development of student capacity to think and reason: An analysis of the relationship between teaching and learning in a reform mathematics project. *Educational Research and Evaluation, 2*(1), 50–80.

Stein, M. K., Smith, M. S, Henningsen, M., & Silver, E. A. (2009). *Implementing standards-based mathematics instruction: A casebook for professional development.* New York, NY: Teachers College Press.

Stigler, J. W., & Hiebert, J. (2009). *The teaching gap: Best ideas from the world's teachers for improving education in the classroom.* New York, NY: Free Press.

St John, M., Fuller, K. A., Houghton, N., Tambe, P., & Evans, T. (2005). *Challenging the gridlock: A study of high schools using research-based curricula to improve mathematics.* Inverness, CA: Inverness Research.

Strauss, V. (2016, May 10). Judge calls evaluation of N.Y. teacher 'arbitrary' and 'capricious' in case against new U.S. secretary of education. *Washington Post.* Retrieved August 10, 2017, from https://www.washingtonpost.com/news/answer-sheet/wp/2016/05/10/judge-calls-evaluation-of-n-y-teacher-arbitrary-and-capricious-in-case-against-new-u-s-secretary-of-education/?utm_term=.1f04c64946b5

Tarr, J. E., Grouws, D. A., Chávez, Ó., & Soria, V. M. (2013). The effects of content organization and curriculum implementation on students' mathematics learning in second-year high school courses. *Journal for Research in Mathematics Education, 44*(4), 683–729.

Toch, T. (1993). The perfect school. *US News & World Report, 114*(1), 46–60.

U.S. Census Bureau (2012). Estimated population and components of change by state: 2010–2012. Retrieved September 6, 2017, from https://factfinder.census.gov/faces/tableservices/jsf/pages/productview.xhtml?src=bkmk

Vacc, N. N. (1993). Implementing the professional standards for teaching mathematics: Questioning in the mathematics classroom. *Arithmetic Teacher, 41*(2), 88–92.

Wiggins, G., & McTigue, D. (2005). *Understanding by design.* Alexandria, VA: Association for Supervision and Curriculum Development.

Wisconsin Act 10. (2011). Retrieved on September 13, 2017, from https://docs.legis.wisconsin.gov/2011/related/acts/10

Yen, W. M., & Ferrara, S. (1997). The Maryland School Performance Assessment Program: Performance assessment with psychometric quality suitable for high stakes usage. *Educational and Psychological Measurement, 57*(1), 60–84.

ABOUT THE AUTHORS

Michael D. Steele is an Associate Professor of Mathematics Education and Chair of the Department of Curriculum and Instruction at the University of Wisconsin-Milwaukee, where he directs the secondary mathematics teacher preparation program. He was previously a member of the Department of Teacher Education at Michigan State University, where he also directed the secondary mathematics teacher preparation program and worked directly with a number of the mathematics faculty at Holt High School. Dr. Steele's work focuses on supporting secondary math teachers in developing mathematical knowledge for teaching, integrating content and pedagogy, through teacher preparation and professional development. He is the principal investigator of the Milwaukee Master Teacher Partnership, a 5-year professional development partnership for high school mathematics and science teachers in Milwaukee Public Schools. He has been a collaborator on NSF-funded projects creating professional development materials focsed on discourse (Mathematics Discourse in Secondary Classrooms), the study of high school curriculum (Mathematical Practices Implementation), and the study of district algebra policy (Learning About New Demands in Schools: Considering Algebra Policy Environments). Dr. Steele has authored a number of research articles focused on mathematical knowledge for teaching and teacher professional development in journals such as *Journal for Research in Mathematics Education, Journal of Mathematics Teacher Education, Journal of Mathematical Behavior, Journal of Research on Leadership Education, Mathematics Teacher Educator, Teachers College Record,* and *Cognition and Instruction.* He is currently the coauthor of two volumes of teacher professional development materials, one focused on reasoning and proving and the other on secondary mathematics classroom discourse (both in press). He is also a coauthor of the NCTM publication Taking Action: Implementing Effective Mathematics Teaching Practices in Grades 6–8 and the associated Principles to Actions Toolkit. He has delivered numerous talks at major mathematics education conferences for researchers and teachers, including invited keynotes at the National Council of Teachers of Mathematics Winter Institute and the National Council of Supervisors of Mathematics Annual Meeting (see CV for complete list). He is currently President-Elect of the Association of Mathematics Teacher Educators.

Craig Huhn is a full time secondary mathematics teacher at Holt High School. In this capacity, he has also serves as a field placement for numerous pre-service teacher candidates from Michigan State University and has previously taught Teacher Education courses at MSU. Over the course of his professional teaching career, Huhn has been an integral part of the mathematics education direction for Holt Public Schools, serving on advisory roles, teacher leader positions, and school and district committees. In the larger mathematics education community, Huhn has presented at various state and national conferences, including the regional National Council of Teachers of Mathematics conference in Boston; Association of Mathematics Teacher Educators in Irvine, CA; and most recently as an invited plenary speaker at the 2015 Psychology of Math Ed- North America conference. In addition, he has twice served as an invited consultant for the next generation of College Preparatory Mathematics curriculum materials in California. In 2013 he was recognized as a finalist for the National Science Foundation's Presidential Award for Excellence in Mathematics and Science Teaching, and in 2015 is a current finalist. His published works include an article in *Ed Leadership* about reconceptualizing the meaning of grades, a chapter in NCTM's *Teachers Engaged in Research* about ways to maximize the learning of students with disabilities in mathematics, a chapter and several responses in *Embracing Reason,* two contributions to SAGE's *Classroom Management* anthology, and two articles in NCTM's *Mathematics Teacher.*